—手術前後飲食自療—

養護好體質

All-round wellness

Pre- / Post-operative healthful recipes

　　芳姐（張佩芳）於香港大學專業進修學院修畢「中醫全科文憑」及修讀中醫全科學士多個課程。自 1996 年起在各大報章及雜誌撰寫食療專欄，作品散見於新晚報、大公報、壹蘋果健康網、加拿大明報周刊「樂在明廚」、及北美洲「品」美食時尚雜誌。

　　芳姐曾任「中華廚藝學院」健康美食班客座講師，現任僱員再培訓局中心、職訓局中心的陪月班導師，及「靚湯工房」榮譽食療顧問。

　　為推廣食療心得，芳姐自 2007 年開始在網上撰寫食療雜誌「芳姐保健湯餸」，深獲網民支持和愛戴。近年開始將其心得著作成書，著有《每天一杯養生茶》、《坐月天書》、《每天一杯養生茶2》、《懷孕天書》、《100 保健湯水・茶飲》、《滋補甜品店》、《女性調補天書》、《強身・治病 孩子飲食全書》、《上班族每天養生法～飲出好體質》。著作迅即再版，成績斐然。

雖然現代醫學昌明，很多疾病都能夠通過手術，甚至「微創」方式進行，減少了風險。但手術無可避免給人體帶來創傷，較大的手術如腹腔手術等更可能會造成病人身體的代謝紊亂、電解質平衡失調、營養不良、貧血、傷口發炎疼痛等，這些手術所帶來的傷害，都會直接影響機體的康復。

病人在手術後，其體質能否儘快恢復，常因個人體質，及因所患疾病的輕重而異，也與術後的調養護理是否恰當有關。因此，外科手術之後，調養至為重要，它直接關係到手術後疾病是否會轉移。因此，合理的調養和護理能讓患者早日康復

手術後進補對康復進度甚有裨益，而許多人的進補觀念常局限在某些特殊保健品和名貴中藥材上，其實人們所需的營養素，最佳來源其實是天然食物。因此，本書採用的，全是天然食物製成的食療，而不是藥療。湯水、茶飲和小菜等都是性質溫和的，但為謹慎起見，每個食療都註明了飲食宜忌、療效和食材介紹等，方便讀者更易了解。

本書因應各種不同的手術，共設計了 70 多款食療湯水、茶飲及小菜，讓讀者可以按病情需要作出選擇，促進病人對營養的吸收，提升免疫功能，強化體質。同時間，透過有益和美味的飲食調理，令病患者深深感受到家人的愛護和關心，因此大大增強了手術後康復的信心，加快痊癒。

目錄

目錄

Contents

手術前

由於做任何手術都會造成身體組織損傷，身體因此需要更多熱量和蛋白質來幫助傷口癒合，及修補受傷組織。所以手術前要較平時多吃一些富含蛋白質類的食物和提高身體熱量的食物，如牛奶、雞蛋、豆類、魚類、肉類等。三餐飲食以蛋白質類及五穀類為主，再視乎情況補充適量的新鮮蔬菜、瓜類、水果。口渴時可飲用一些米湯、豆漿及乳製品，除了補充水分，也可同時補充蛋白質與熱量。

新鮮蔬果所含熱量少，且易產生飽足感，若食欲不好，不宜進食太多。手術前要採取高熱量、高蛋白質飲食以確保營養足夠，尤其是對一些患上惡性腫瘤需要電療、化療的病人，高蛋白質飲食可避免體重減輕及體內組織耗損。

對食欲不振的病人，可依其食欲及體能狀況每2至3小時進食一次，實行少食多餐。攝取高蛋白質及熱量固然重要，但家禽類、豬肉、羊肉等必須去皮、去肥膏，以免吸收過多動物性油脂對健康不利；甜食雖然能提高熱量，但不建議多食；含反式脂肪的蛋糕、點心則應盡量少食。手術前的調養，把身體的狀況調到最佳狀態，更能提高手術的成效。

雖然手術前要補充多些營養，但不建議未經醫生指示下胡亂服用保健產品，因為不少保健產品中各種成分病人未必完全了解，當中亦可能含有一些遏制血小板凝血功能的物質，或行氣活血的中藥材，致令手術期間難以控制出血的情況。

手術後

除了一些局部淺表的手術，大部分手術前都需要禁食。局部淺表手術失血不會太多，這類體積小及位置淺，只需局部麻醉下可施行手術切除，對身體氣血損害較小，只需手術後對傷口照顧得好就可以。手術後亦只需服用一些能促進傷口癒合的食療就成。

創口較大的手術，由於創傷及麻醉用藥的關係，需要留意的事情就會較多。病人表面傷口未癒合之前，以及脾胃功能未完全恢復，不宜給大補氣血的湯水供病人服用。有些胃腸道手術，胃腸蠕動未恢復前，腸道處於低功能狀態，必須禁食。就算可以服食流質或半流質食物，亦需注意飲食要清淡富營養，即使食物煮到很爛，也須等手術後 8 至 10 天，依照醫護人員指示才能酌情給予。

有些病人手術後覺得疼痛和疲倦，因而不思飲食，尤其是很多癌症病人手術後會出現食欲不振，或因手術導致的副作用而無法正常進食，往往要一兩天後才可飲少量飲料，以防脫水。麻醉手術會影響腸道蠕動，補充水分亦要注意只能小口、小口喝水，必須以漸進方式進行。

辛辣食物如大蒜，本來有益，但對心臟病手術者未必好，因大蒜能抑制某些藥物的有效成分，甚至與某些藥物中的成分發生化學反應而產生毒素。對於手術後要服用薄血丸人士，更需遵照醫生指示，少吃維他命 K 的食物。因為維他命 K 與薄血藥有抗衡作用，降低薄血藥之療效。深綠色葉蔬菜如韭菜、芥蘭、莧菜、菠菜、通菜、豆苗、枸杞葉等，多含有量高之維他命 K。此外，動物肝臟、黑木耳、牛油果、木瓜及一些具活血祛瘀作用的中藥材及維他命丸、酒品等亦盡量少用，以免影響薄血藥的療效。

手術後不宜吃油膩、滋補及燥熱的食物，因為身體一開始可能無法適應油膩和滋補的食物，故需避免攝取，以免脾胃無法負荷。待病人手術後情況穩定，沒有併發症發生才開始調補，這樣調養才能事半功倍。

手
術
前
後
・
飲
食
的
謬
誤
及
建
議

手術前後要考慮的，第一是脾胃運化能力，第二是病情，第三是病患者的體質。手術後的飲食重清淡不肥膩，容易消化，不會妨礙脾胃功能，以使患者有良好消化能力而能早日康復。所以太過肥膩、多油、煎炸、刺激辛熱及生冷寒涼的食物均不宜吃。

有些人平日習慣吃濃味的食物，家人為了讓病人增加食欲，會特意烹調一些濃口味的食物供患者食用，但過甜、過鹹、太酸和辛辣食物易刺激胃腸黏膜，或加重腎臟、肝臟負擔，或引致腹瀉及胃腸氣脹症狀，均不利病情康復，故手術前後飲食宜以清淡為主。

中醫一向認為，筍、蝦蟹、鵝肉、公雞肉、豬頭肉（包括豬頸肉）、牛肉、無鱗魚、芒果、菠蘿等均屬發物，易誘發炎症，故手術後必須禁食發物。有人對這種說法有異議，因為他們說吃了也沒有任何不適，但事實上所謂的「發物」，是對一些本身飲食不節，肥胖濕多，身體四肢容易腫脹，體內積濕熱，易困倦乏力，皮膚長暗瘡濕疹，中醫認為屬「濕熱內蘊」者，這類人吃了發物，體內的濕熱蘊毒便容易發散出來。而事實上，手術後兩週，即使已拆線，這段時間身體抵抗力還是很弱，炎症發生的危險依然存在，故為病情快些康復，這類「發物」還是少食為宜。

有些人說手術前後吃薑，會因過於行氣而令手術出現出血情況，或令傷口長肉芽。其實有大學研究顯示：病人在手術前一小時吃些少薑，術後出現嘔吐的機會，比沒有吃薑的低三成半，原因是薑的止嘔作用可改善因麻醉藥而出現的手術後噁心和嘔吐的後遺症。至於產後煮湯加薑，會令傷口長肉芽，這點其實未有科學驗證，為免病患者心情忐忑不安，建議生薑改用陳皮就可以了。然而像人參、丹參、田七、川芎一類行氣活血的中藥材，並不建議於手術前兩、三天或手術後一週內服食，以免造成手術期間及手術後難以控制的出血情況。

很多病患者會於手術前後服食過多營養品和補品，其實這樣會增加肝臟的負擔，而過量的營養品和補品不但不易被肝臟所分解，還會使肝臟處於超負荷運轉，不利於身體的恢復。其實患者只要注意飲食營養均衡，多採用天然食材，本着清淡易消化吸收的原則作合理調補，自然有利於身體的康復。

010

手術甦醒後——可以飲的清流質飲食

清流飲食的特點是完全無渣，不產氣、不刺激腸胃道蠕動，以供應水分為主，在室溫或體溫時為清澈液體的流質飲食。例如水、米湯、果汁、蜂蜜水、運動飲料等。

清流質飲食可提供水分、部分電解質及少許熱量，並可減少糞便及渣滓至最少量，以幫助胃腸道功能的恢復，使病患者盡快正常飲食。

手術後——可以進食的半流質飲食

半流質飲食是將固體食物經由剁碎、絞細等方式處理，加入飲料或湯汁，調製成稍加咀嚼即可吞嚥之飲食，如：稀粥、小米粥、蛋花湯、雜菜湯、米湯、軟爛麵條等。

半流質飲食的目的，是使吞嚥或咀嚼固體食物稍有困難的病者，仍能得到足夠的營養。通常適用對象是一些無牙咀嚼、吞嚥稍有困難、胃炎、消化不良、急性熱病期的病者。

半流質飲食宜少量多餐，營養分配均衡，食物的選擇以質地細軟，易消化為原則。過老或有筋的肉類、粗糙的蔬果、堅果、豆類及油炸食物均不宜食用。

手術後——澱粉質主食

手術後病人若無不適，則進展為溫和飲食，以澱粉質為主食，例：米飯、通粉、麵條、米綠等三餐為正餐。

澱粉質主食的食材不要經油炸及過硬，配料盡量剁細處理。初進食時盡量採用軟質食物，如豆腐、低纖維之嫩葉及瓜果類，肉類宜先去皮、去肥膏。漸進採用低渣、體積小、易消化、溫和性的食材為主。

電療、化療期的飲食調理

現時治癌三法是外科手術治療、放射治療（即電療）、化學藥物治療，其目的都是消除癌腫。放射治療是用放射性元素蛻變產生電療輻射在治療上的應用，這種治療方法，可抑制和破壞某些癌細胞，使腫塊很快縮小或消失。但由於放射線對癌細胞和正常細胞同時有破壞作用，故在治療癌腫的同時，正常細胞亦會受到一定損害，如照射過的皮膚出現萎縮、變薄和變白、色素沉着和毛細血管擴張；患者或會有頭暈眼花、煩躁疲乏、嗜睡或失眠、血中白細胞總數下降與及血小板減少等全身反應，或噁心嘔吐、食欲減退等腸胃反應。

化學藥物治療，對癌細胞有一定的抑殺作用，但化學治療藥物常因缺乏對癌細胞特異性的選擇作用，而會引起全身性反應，因此，化療藥物的毒性作用和不良反應對機體正常細胞或某系統往往會產生不同程度的損害。如對骨髓細胞、胃腸道黏膜上皮細胞、生殖細胞、毛髮等損害較為明顯。

　　既然電療、化療對癌症患者有這樣多的損害，因此對癌症患者的支持十分重要，要根據患者的特殊需求提供良好的營養，有利於治療和康復。電療期間因產生放射反應，患者常有頭暈、煩躁、失眠、口苦、噁心或嘔吐等，若兼有小便黃大便秘結、口乾渴飲等情況，乃熱傷肺胃，引致陰虛內熱，飲食調理宜用清肺養胃、滋潤生津等法，可選用甘寒清淡的食物，如雪梨、竹蔗、馬蹄、冬瓜、薏米、綠豆、海藻、雪耳、百合等作調理。

　　化療期間，往往胃腸反應較為明顯。由於胃腸反應，食欲不振，患者常因此而顯得異常虛弱。飲食方面宜選用高蛋白、高熱量、含豐富維生素而又易於消化吸收的食物如：魚肉、瘦肉、雞胸肉、魚丸、肉丸、海參、花膠、米糊、豆腐，豆漿、雞蛋、雪耳、猴頭菇、木耳、杞子、黑芝麻等；新鮮的蔬菜、瓜果如：蘆筍、番茄、薯仔、紅蘿蔔、菠菜、蓮藕、鮮淮山、合掌瓜、節瓜、苦瓜、青瓜、南瓜、絲瓜、葡萄、藍莓、蘋果、奇異果、香蕉、檸檬、橘橙、百香果等，需適量食用。同時宜少量多餐，以增進食欲，減少不良反應。

南瓜營養湯

Nutritious pumpkin soup

▌材料（2～3人量）

南瓜	150 克
洋葱	1 個
薯仔	1 個
菠菜	100 克
紅腰豆	60 克

▌調味料

海鹽	1/4 茶匙

▌做法

1. 南瓜、薯仔去皮，洗淨切件；洋葱去衣，切件；紅腰豆浸洗；菠菜去根，洗淨切段。

2. 燒熱 1 公升水，放入南瓜、洋葱、薯仔及紅腰豆，煮 1 小時，加入菠菜及調味，再煮 5 分鐘即可連湯料同食。

食療功效

補中益氣、消炎解毒。

飲食宜忌

本品營養豐富，老少皆宜。對手術前或手術後煩躁口渴、高血壓、眼底出血、脾胃虛弱、腸胃失調者有益。一般人士皆可飲用，但洋葱、菠菜具有抗凝血物質，故手術前 2-3 天及手術後一週不要吃太多洋葱和菠菜。

認識主料

紅腰豆

紅腰豆是豆類中營養較豐富的一種，有補血、增強免疫力，幫助細胞修補及防衰老等功效；並能降低膽固醇及血糖。一定要煮到夠熟才可食，否則易導致過敏反應。痛風症患者不宜。

南瓜

南瓜能補中益氣、消炎止痛。南瓜含豐富的果膠，有助排走身體有害物質；鮮榨南瓜汁亦能加快腎結石和膀胱結石的溶解。

粉葛赤扁豆大棗湯

Kudzu red date soup with
small red beans and hyacinth beans

材 料 （ 2 ~ 3 人 量 ）

粉葛	250 克
赤小豆	30 克
扁豆	30 克
冬菇	4 朵
陳皮	1 塊
紅棗	6 粒

調 味 料

海鹽	1/4 茶匙

做 法

1. 粉葛撕去外皮，洗淨切塊；冬菇浸軟，去蒂；赤小豆、扁豆、陳皮浸洗；紅棗去核。
2. 全部材料放入煲內，用 1.2 公升水煮個半小時，調味即可供服。

食療功效

止渴除煩、疏通血管。

飲食宜忌

本品香濃可口，老少皆宜。適合手術前或手術後肩頸肌肉緊張，心情煩躁、鬱悶氣促、精神不振者服用。一般人士皆可服。

手術前後營養素菜湯

認識主料

粉葛

粉葛能止渴除煩，提高肝細胞再生等能力。粉葛塊頭大，表面凹凸，削皮不易，故最好用小刀壓住表皮頂部，一塊塊撕下來；而縱向破開較橫切省力很多，破開後才細切成塊。

赤小豆

赤小豆能健脾利濕，散瘀，解毒。可用於水腫、腳氣、產後缺乳、腹瀉、小便不利、痔瘡等。赤小豆一般人群可食，因含嘌呤少，且能抑制尿酸，故痛風患者都可以食用。夜尿多者宜少食。

牛蒡排毒湯

Burdock detox soup

材料（2～3人量）

鮮牛蒡	150 克
小香菇	6 朵
紅蘿蔔	60 克
魔芋絲	100 克
菠菜	150 克
素上湯	2 碗

調味料

海鹽	1/4 茶匙

做法

1. 牛蒡連皮洗擦乾淨，切片；香菇浸軟，去蒂；紅蘿蔔去皮，切塊；菠菜去根，洗淨切段；

2. 煮滾素上湯及 500 毫升水，加入牛蒡、香菇、紅蘿蔔煮 40 分鐘，最後加入魔芋絲、菠菜，再煮 10 分鐘，調味即可連湯料同食。

認識主料

魔芋

魔芋是蒟蒻芋加工製成。它含有豐富的纖維素，熱量極低，有「胃腸清道夫」之稱。因為魔芋吸水力很強，容易產生飽脹感，也經常被視為減肥食品，適合便秘、高血脂、糖尿病和防治胃癌、腸癌人士食用。但脾胃虛寒者忌食。

牛蒡

牛蒡能清熱解毒，疏風利咽。常用於風熱感冒、咳嗽，咽喉腫痛、瘡癤腫痛、濕疹等症。它更有防癌、抗癌等作用；對高血脂、糖尿病、便秘、老年血管硬化等病症有一定療效。但脾胃虛寒者宜少食。

食療功效

清熱排毒、養血潤腸。

飲食宜忌

這湯有助消脂減肥、排毒養顏。適合肥胖、高血壓、糖尿病及中風人士手術前及手術後飲用。但脾胃虛寒及便溏者忌服。手術前或手術後需服用薄血藥者不宜用菠菜，可用椰菜代替。

小貼士

可以善用廚餘製作素上湯，例如棄用的冬菇腳、菠菜根、用剩的紅蘿蔔、青瓜等，只要加些提鮮的黃豆、海藻同煮，熬上 2 小時，就是鮮甜美味的素上湯。

白果腐竹馬蹄黃豆湯

Soybean soup with gingkoes,
dried tofu skin and water chestnuts

手術前後營養素菜湯

▌材 料 （ 2 ～ 3 人 量 ）

白果	15 粒
腐竹	1 孖
馬蹄	6 粒
黃豆	60 克
陳皮	3 克

▌調 味 料

海鹽	1/4 茶匙

▌做 法

1. 白果去殼、去衣及去芯；腐竹沖洗；馬蹄去皮，洗淨；黃豆、陳皮浸洗。
2. 將全部材料用 1 公升水煮滾，改用文火煮 45 分鐘，調味即可連湯料同食。

▌食療功效

健脾開胃、滋陰補腎。

▌飲食宜忌

本品不寒不燥，營養豐富，容易消化及吸收。對手術前及手術後肺腎虛弱，喘咳多痰者有益。一般人群可食。

認識主料

黃豆

黃豆含卵磷脂，可增強神經機能和活力。同時黃豆能提高精力、降低血脂、預防癌症、提升免疫功能作用。但痛風患者及甲狀腺機能低下患者不宜。

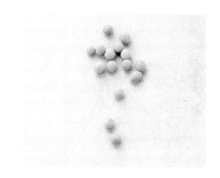

白果

白果能斂肺定喘，止帶濁，縮小便。用於痰多喘咳、帶下白濁、遺尿及尿頻。但邪實痰多者不宜食，白果有小毒，不宜多食常食；五歲以下小兒忌食白果。成人每次食用以 15 粒為限。

▌材 料（2～3 人量）

蓮藕	250 克
紅蘿蔔	1 條
花生	50 克
冬菇	4 朵
生薑	3 片
紅棗	6 粒

▌調 味 料

海鹽	1/4 茶匙

▌做 法

1. 蓮藕、紅蘿蔔去皮，洗淨切塊；花生、冬菇浸洗，冬菇去蒂；紅棗去核。
2. 燒熱 1 公升水，加入全部材料煮 1 小時，調味即可連湯料同食。

食療功效

補血強身、消除疲勞。

飲食宜忌

本品清甜美味，適合手術前或手術後氣血虛弱、面色蒼白、食欲不振、神疲乏力者服用。但消化不良者不宜多食湯料。

認識主料

蓮藕

蓮藕有七孔和九孔，藕孔越多越脆嫩香甜，七孔藕澱粉含量較高，水分少，糯而不脆，較適宜煮湯。蓮藕忌鐵器，最好用瓦煲、瓷煲或康寧煲來煮湯。

蓮藕紅蘿蔔花生湯

Lotus root soup with carrot and peanuts

海藻豆腐金菇湯

Seaweed tofu soup with enokitake mushrooms

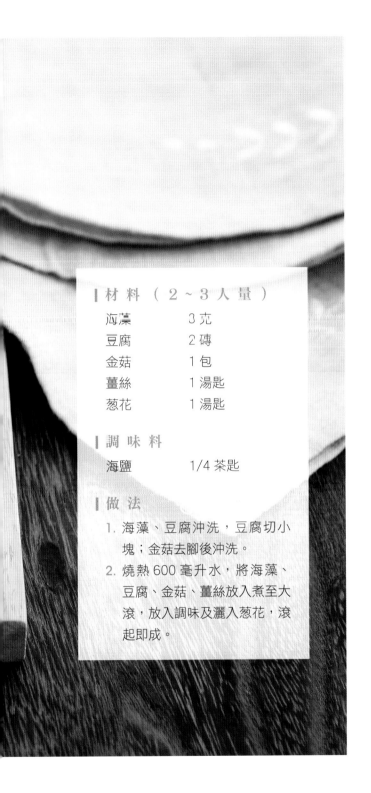

材料（2~3人量）

海藻	3克
豆腐	2磚
金菇	1包
薑絲	1湯匙
蔥花	1湯匙

調味料

海鹽	1/4 茶匙

做法

1. 海藻、豆腐沖洗，豆腐切小塊；金菇去腳後沖洗。
2. 燒熱 600 毫升水，將海藻、豆腐、金菇、薑絲放入煮至大滾，放入調味及灑入蔥花，滾起即成。

食療功效

清熱散結、利水降壓。

飲食宜忌

本品鮮味香滑，老少皆宜。適合甲狀腺腫、頸淋巴結核、高血壓、高血脂、冠心病者手術前或手術後服用，癌症患者電療、化療期間服用亦佳。但脾胃虛寒、氣血兩虧者不宜多食。

認識主料

海藻

海藻含有二十餘種人體必需的胺基酸，除可幫助降血壓及降膽固醇外，尚有助排除體內放射性物質。故很適合電療、化療病者食用。

手術前後營養素菜湯

蘆筍雪耳芙蓉羹

Egg drop soup with asparagus and white fungus

手術前後營養素菜湯

材料（2～3人量）		調味料	
蘆筍	200克	海鹽	1/4 茶匙
雪耳	6克		
杞子	3克		
蛋白	2個		
素上湯	3碗		
生粉水	1湯匙		

▌做 法

1. 蘆筍洗淨，將硬皮削去，切粒；雪耳浸軟，去蒂後切碎；杞子浸洗；蛋白打散。

2. 燒熱素上湯，放入雪耳煮20分鐘，加入蘆筍、杞子煮5分鐘，調入生粉水，邊倒邊攪拌，最後加入打勻的蛋白及調味，待蛋白浮起即可熄火供食。

食療功效

利尿降壓、消除疲勞。

飲食宜忌

本品味美，老少皆宜。適合高血壓、冠心病、胃腸病、貧血、關節炎、浮腫、神經炎、肥胖、癌症人士手術前或手術後服用。但痛風症患者不宜。

認識主料

蘆筍

蘆筍能治療貧血、疲勞症、關節腫痛、神經炎等症，還有祛濕消腫功效。蘆筍含微量元素硒，能增強病人對癌症的抵抗力，但若用於治療腫瘤，要連續不斷食用，一直至腫瘤消除，或被控制為止，否則療效不顯。但痛風患者不宜食。

蘆筍含有的維生素A多聚在筍尖上，選購時留意不是越粗壯越好，反而粗幼度適中、質地硬實、葉片緊密及顏色青翠的食療功效更佳。

雪耳

雪耳能補肺益氣、養陰潤燥。對肺虛咳嗽、痰中帶血、崩漏、大便秘結、高血壓、血管硬化等症有幫助。對陰虛火旺虛不受補者及癌症電療、化療者均有裨益。但外感風寒及出血症病人慎服。

椰子黃耳杏仁湯

Coconut soup with yellow ear and almonds

材料（2～3 人量）

印度椰子	1 個
黃耳	10 克
甜杏仁	20 克
花生	30 克
杞子	3 克
紅棗	6 粒

調味料

海鹽	1/4 茶匙

做法

1. 印度椰子切塊，沖洗淨；黃耳用清水浸軟，去蒂，剪碎；杏仁沖洗；花生、杞子分別浸洗；紅棗去核。
2. 燒熱 1 公升水，放入全部材料，煮約 1 小時，調味即成。

滋補強身、養顏潤肺。

本品滋補而不膩滯，老少皆宜。適合肺弱咳喘、胃及十二指腸潰瘍、津液不足、面色萎黃、皮膚乾皺者手術前或手術後服用。但感冒發熱者不宜。

手術前後營養素菜湯

印度椰子

印度椰子屬高蛋白質、低熱量生果，含
量量的維生素B，對乾咳、改善尿道炎、
心臟功能及中和胃酸等均有幫助。但印
度椰子較易霉變，購買時最好要求商販
破開才買。

黃耳

黃耳有清心補腦、延緩衰老、化痰止咳
等功效。並能提高機體代謝機能，抑
制腫瘤細胞生長，並有預防脂肪肝作
用。但痰濕體質者不宜。

手術前後營養素菜湯

竹笙金針雪耳湯

Bamboo fungus soup with day lily and white fungus

材料（2～3人量）

竹笙	6條
金針	6克
雪耳	5克
杞子	3克
急凍粟米粒	2湯匙
去殼栗子	10粒

調味料

海鹽	1/4茶匙

做法

1. 竹笙、金針、雪耳浸軟，分別去蒂，金針打結；杞子浸洗；急凍粟粒沖洗解凍；栗子投入開水中煮片刻，去衣。
2. 燒熱800毫升水，加入全部材料煮30分鐘，調味，即可連湯料同食。

食療功效

養血止血、解鬱安神。

飲食宜忌

本品清甜而不肥膩，對手術前及手術後虛熱又有燥火、虛煩失眠、口苦咽乾、視瞢矇矓、精神抑鬱、肥胖、高血壓者很有幫助。但由於雪耳含有一些抗凝血物質，故手術前2-3天和手術後一週都不宜吃雪耳，可用香菇代替。

認識主料

竹笙

竹笙以色澤淺黃、長短均勻、質地細軟、氣味清香者為佳品。竹笙具有止痛補氣，降血壓、降膽固醇等功效，由於竹笙甚為「刮油」，故具有減少腹壁脂肪積聚的作用。

手術前後營養素菜湯

■ 材料（2～3 人量）

合掌瓜	2 個
腰果	50 克
蘑菇	60 克
牛蒡	2 片

■ 調 味 料

海鹽	1/4 茶匙

■ 做 法

1. 合掌瓜去皮，切塊；腰果沖洗；蘑菇去蒂，沖洗。
2. 燒熱 800 毫升水，放入全部材料，煮約 30 分鐘，調味即可連湯料同食。

合掌瓜腰果蘑菇湯

Button mushroom soup with chayote and cashew nuts

———— 食療功效 ————

舒肝理氣、健腦益智。

———— 飲食宜忌 ————

本品營養豐富又美味，老少皆宜。對手
術前或手術後肝氣鬱結、肝胃氣痛、高
血壓、高膽固醇及腦力衰退者有益。一
般人士皆可服。

———— 認識主料 ————

合掌瓜

合掌瓜在瓜類中營養較全面，能增強人
體免疫力，並有利尿作用，能擴張血管、
降血壓。它含鋅量高，對兒童智力發展
有幫助。此外，對男女因營養不良引致
的不育亦有幫助。合掌瓜以細嫩、飽滿、
果皮表面縱溝較淺、皮色青翠者為佳。
一般人士皆可食用。

腰果

腰果含豐富維生素 A、B 及蛋白質等營養
素，能排毒養顏、潤腸通便、延緩衰老。但
過敏體質人士吃了腰果會引起強烈過敏反
應，甚至休克；故未吃過腰果者先吃兩
粒，待 10 分鐘後如有流口水、嘴內刺癢
情況就不要再吃。

青木瓜花生響螺肉湯

Conch soup with green papaya and peanuts

▌材 料（2～3 人量）

青木瓜	1 小個
花生	50 克
急凍響螺	100 克
無花果	3 粒
瘦肉	150 克

▌調 味 料

海鹽	1/4 茶匙

做法

1. 青木瓜去皮，去核，切塊；花生浸洗；急凍響螺解凍，汆水；瘦肉切片，汆水。
2. 將全部材料用1公升水煮1小時，調味即可連湯料同食。

食療功效

平肝和胃、滋陰潤肺。

飲食宜忌

本湯鮮甜美味，老少皆宜。適合胃炎、胃潰瘍、十二指腸潰瘍、虛熱煩悶人士手術前或手術後服用。一般人士皆宜服用，但孕婦不宜食木瓜，因會令子宮收縮。可用響螺乾代替急凍響螺。

青木瓜

未成熟的青木瓜含一種乳汁，能保護胃腸黏膜，故對患有消化系統疾病的人士很有助益。木瓜和肉類一起烹煮，不僅能增加鮮味，而且能使肉類很快熟爛。

響螺乾

響螺煲湯宜保留其掩蓋同用，因為響螺的掩含鈣、鈉、鉀、鐵等豐富礦物質及多種維生素，有較高的營養價值。而未成熟的青木瓜含一種乳汁，能保護胃腸黏膜，故對消化系統疾病很有益。木瓜和肉類一起烹煮，不僅能增加鮮味，而且能使肉類很快熟爛。

鮮淮山鮮石斛燉海參

Double-steamed sea cucumber soup
with fresh yam and Shi Hu

材料（2～3 人量）

鮮淮山	150 克
鮮石斛	50 克
雪耳	5 克
雞胸肉	1 塊
浸發海參	2 條
南棗	4 粒
生薑	2 片

調味料

海鹽	1/4 茶匙

做法

1. 鮮淮山去皮，洗淨，切塊；鮮石斛洗淨，切段；雪耳浸軟，去蒂；雞胸肉切塊，汆水；海參切塊，汆水。
2. 將全部材料放入燉盅內，注入開水 500 毫升，隔水燉 3 小時，調味即成。

食療功效

益胃生津、養血潤燥。

飲食宜忌

本品滋補美味，老少皆宜。適合手術前或手術後或癌症患者電療、化療後出現津液不足、氣虛血弱、神疲乏力、視力減退、大便秘結者服用。但外感發熱，痰濕壅滯或便溏者忌服。

認識主料

鮮石斛

鮮石斛有青皮、紫皮兩種，以紫色皮的鮮石斛療效較佳。鮮品洗淨入口細嚼，味甘而微黏，清新爽口，常食有助提高身體免疫能力、降血糖、抗腫瘤。鮮石斛在一些中藥海味店偶有售，如買不到，用乾品 6 克代替。

洋蔥番茄牛肉湯

Beef soup with onion and tomato

材料（2～3人量）

洋葱	1 個
番茄	3 個
牛肉	300 克
生薑	2 片

調味料

海鹽	1/4 茶匙

做法

1. 番茄去皮，洗淨，切塊；洋葱去衣，切片；牛肉切片，汆水。
2. 全部材料用 800 毫升水煮半小時，調味即成。

食療功效

健脾益胃、強壯體質。

飲食宜忌

本品能增強體力及免疫力，老少皆宜。適合高血壓、糖尿病、冠心病、肥胖症、癌症者手術前或手術後服用。但牛肉屬發物，患有濕疹、皮膚病患者宜少食牛肉；而洋葱亦含有抗凝血物質，故手術前2-3天及手術後一週不要吃洋葱。

認識主料

洋葱

洋葱有祛風發汗、解毒消腫的功效。適合感冒風寒、頭痛鼻塞、中風、面目浮腫、痢疾等、瘡腫等。洋葱能溶解血栓，並能降低膽固醇、防動脈粥樣硬化。但肺胃有熱、陰虛、有出血症者宜少食。但切洋葱時強烈刺激的氣體皆令人流淚，可將洋葱浸在水中切，使氣味溶入水中不散發出來。

牛肉

牛肉有補脾胃、益氣血、強筋骨的功效。對體虛乏力、筋骨痠軟、氣虛自汗者有益。牛肉的蛋白質含量高豬肉兩倍，含鐵量也豐富，對手術後失血過多者及修復組織和創傷很適合。牛肉屬發物，有皮膚濕疹、瘡毒者忌食。外感未清亦不宜食用。

節瓜海藻魚片湯

Grass carp soup with
Chinese marrow and spirulina

手術前後調養體質濃湯

▌材 料 （ 2～3 人 量 ）

節瓜	1 條
螺旋海藻	5 克
鯇魚脊肉	1 條
薑絲	半湯匙

▌調 味 料

海鹽	1/4 茶匙

▌做 法

1. 節瓜刮皮，切塊；海藻沖洗；鯇魚肉洗淨，切薄片。
2. 燒熱 800 毫升水，將節瓜、薑絲煮 15 分鐘，加入海藻、鯇魚肉再煮 10 分鐘，調味即成。

補益氣血、清熱散結。

飲食宜忌

本品鮮甜味美，老少皆宜。適合氣血不足、營養不良、高血壓、高血脂、糖尿病、缺鐵性貧血及患有腫瘤的人士手術前或手術後服用。一般人士皆可飲用。

認識主料

螺旋海藻

海藻是生長在海中的藻類，種類繁多，有大葉海藻、小葉海藻等。螺旋藻其實不屬於海藻，而是屬於湖藻，著名螺旋藻產地有非洲的旦愿湖、中國雲南麗江的程海湖等。而在海裏生長的螺旋藻都是人工養殖的，故稱為螺旋海藻。各類海藻中豐富的葉綠素可以協助人體清除腸毒素，保護肝臟細胞免被毒素干擾，加速身體排毒，減輕代謝廢物對腎臟的負擔。螺旋海藻更能減輕癌症放療、化療的毒副反應。並能提高免疫力、降低血脂。對高血脂、缺鐵性貧血、糖尿病、營養不良、病後體虛等很有幫助。是病後、手術前及手術後的食用佳品。一般人群皆可服，脾胃虛寒者宜少食。

手術前後調養體質濃湯

蟲草孢子頭花膠燉水鴨湯

—
手術前後調養體質濃湯
—

Double-steamed teal soup with cordyceps spores and fish maw

材料（2~3人量）

蟲草孢子頭	6克
浸發花膠	150克
杞子	5克
圓肉	10粒
冰鮮水鴨	1隻
生薑	2片

調味料

海鹽	1/4 茶匙

做法

1. 蟲草孢子頭、杞子、圓肉浸洗；水鴨劏洗淨，斬件後與浸發花膠同汆水。
2. 全部材料放入燉盅內，注入3碗開水，隔水燉3小時，調味即成。

食療功效

滋陰補虛、益氣養血。

飲食宜忌

本品清香味美，老少皆宜。適合陰虛內熱、煩躁不安、糖尿病、高血壓、肺弱咳喘及各種癌症手術前或手術後服用。但外感發熱者及消化不良者不宜。

認識主料

蟲草花

蟲草花有滋肺補腎、護肝、抗氧化、防衰老等作用。對肺腎兩虛、精氣不足、咳嗽氣短、腰膝酸軟等症有幫助。蟲草花的孢子頭越大療效越佳。但陰虛火旺者忌食。

水鴨

水鴨能補虛暖胃、強筋壯骨、活血行氣，且不寒不燥，是手術後調養佳品。傳統的蟲草燉水鴨是珍貴的滋陰補腎強壯劑，但冬蟲草售價太貴，可用蟲草花代替，同樣具滋補作用。水鴨適合作為手術後、或癌症化療後的食療。但水鴨肉滋膩，患外感、腸炎、慢性腹瀉者忌食。

花膠

花膠功能滋腎益精、養血退虛熱。花膠有分雌雄，肚姆呈「環狀」紋，較厚身但煮時易「瀉水」及黏牙；肚公呈「人」字紋，薄身肉爽，煮起來不易溶化。

核桃蓮子珍珠肉湯

Pearl clam soup with walnuts and lotus seeds

材料（2～3人量）

核桃肉	30 克
蓮子	20 克
芡實	20 克
杞子	5 克
珍珠肉	3 個
瘦肉	250 克

調味料

海鹽	1/4 茶匙

做法

1. 瘦肉切片，汆水；珍珠肉汆水；其餘材料浸洗。
2. 將全部材料用 1 公升水煮個半小時，調味即成。

食療功效

健脾補腎、補肝明目。

飲食宜忌

本品香濃味美,老少皆宜。適合肝腎虧虛、視力減退、精神不振、夜尿多者手術前或手術後服用。但外感發熱者不宜。

認識主料

珍珠肉

珍珠肉是珍珠蚌的肉曬乾而成,屬名貴海味,有安神定驚,明目消翳,解毒生肌的功效。對驚悸失眠、驚風癲癇、視力衰退、夜尿多、瘡瘍不斂者有療效。但痛風患者不宜。

珍珠肉以澳洲出產的特別肥美及鮮甜,其色澤微紅有光澤,品質最佳。

豬肉

豬肉滋陰潤燥、滋養健身。對熱病傷津、消渴瘦弱、燥咳、便秘者皆有益,但豬肉多食助濕生痰、蘊濕。外感風寒者忌食。

豆腐泥鰍魚湯

Pond loach soup with tofu

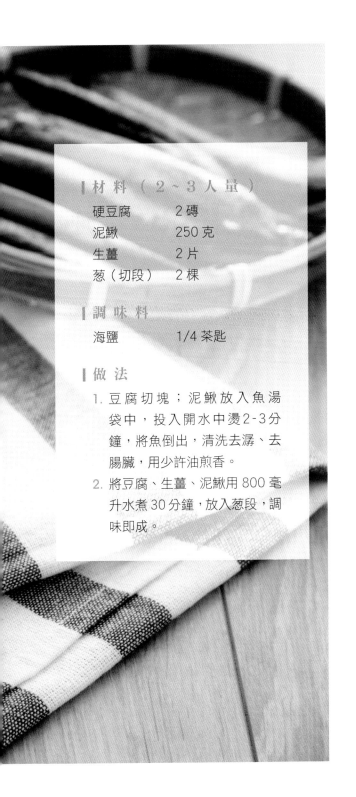

材料（2~3人量）

硬豆腐	2 磚
泥鰍	250 克
生薑	2 片
葱（切段）	2 棵

調味料

海鹽	1/4 茶匙

做法

1. 豆腐切塊；泥鰍放入魚湯袋中，投入開水中燙 2-3 分鐘，將魚倒出，清洗去潺、去腸臟，用少許油煎香。
2. 將豆腐、生薑、泥鰍用 800 毫升水煮 30 分鐘，放入葱段，調味即成。

食療功效

健脾益氣、除濕退黃。

飲食宜忌

本品鮮甜，老少皆宜。對肝炎、肝癌、肝硬化腹水患者手術前及手術後很適合。但脾胃虛寒者要加 1 茶匙胡椒粒同煮。

認識主料

泥鰍

泥鰍有「水中人參」之稱，有補脾胃、利水退黃的功效，同時能促進黃疸消退和轉氨酶下降，有良好的護肝作用。泥鰍只能吃鮮活的，但泥鰍細條又黏滑，劏洗不易，故可用這個入魚湯袋的方法，將魚先行處理了才去潺劏洗烹煮。

Quail soup with dried cuttlefish and red kidney beans

墨魚乾紅腰豆鵪鶉湯

材料（2～3人量）

紅腰豆	30 克
墨魚乾	60 克
紅棗	4 粒
鵪鶉	2 隻
生薑	3 片

調味料

海鹽	1/4 茶匙

做法

1. 紅腰豆浸洗；墨魚乾浸洗，和劏洗淨的鵪鶉一齊汆水；紅棗去核。
2. 將全部材料用 1.2 公升水煮個半小時，調味即成。

食療功效

補益氣血、調養肝腎。

飲食宜忌

本品鮮味可口，老少可飲。對血虛頭髮早白、頭暈目眩、心悸，尤其是電療後出現白細胞減少者有益。但有外感發熱及皮膚過敏者不宜。

認識主料

墨魚乾

墨魚乾能補肝血、滋腎陰，同時對婦女崩漏、來經量多很有幫助，且有催乳和安胎的功效，故為婦科食療佳品。其所含的多肽，有抗病毒、抗輻射線的作用。但痛風、高血壓、心血管疾病、腎臟病、糖尿病患者和皮膚易過敏者忌食。

墨魚乾以乾身、香氣足、無受潮者為佳。

鵪鶉

鵪鶉素有「動物人參」之稱，有滋補五臟、厚腸止痢、祛濕通痹的功效。鵪鶉對貧血、營養不良、神經衰弱、氣管炎、心臟病、高血壓、肺結核、小兒疳積、月經不調病症都有理想療效。但外感發熱者不宜。

金針菠菜豬肝湯

Pork liver soup with day lily flowers and spinach

▊ 材 料 （ 2～3 人量 ）

金針	6 克
菠菜	200 克
杞子	4 克
豬肝	250 克

▊ 調 味 料

海鹽	1/4 茶匙

▊ 做 法

1. 金針浸軟，打成結；菠菜去根，洗淨切段；杞子浸洗；豬肝清洗後切片，汆水。
2. 全部材料用 600 毫升水煮滾，煮 15 分鐘，調味，即可連湯料同食。

食療功效

養血潤燥、鎮靜安神。

飲食宜忌

本品清香，老少可飲。適合肝炎、神經衰弱、乳房脹痛、夜盲症、小便不通、大便下血人士手術前或手術後服用。但菠菜含抗凝血物質，服薄血丸患者、痛風患者及腹瀉便溏者不宜。

認識主料

金針

金針有極佳的安神作用，對神經衰弱者有鎮靜催眠作用，唯金針含秋水仙鹼，而且鮮品的含量非常多，過食易引起嘔吐、腹瀉等食物中毒症狀，食用乾品較安全。

豬肝

豬肝中富含蛋白質、卵磷脂和多種微量元素，有補眼及促進兒童發育功效。

材料（2～3人量）

蓮子	30 克
茯神	20 克
桂圓肉	15 克
芡實	30 克
排骨	250 克

調味料

海鹽	1/4 茶匙

做法

1. 蓮子、茯神、芡實用清水浸洗；圓肉沖洗；排骨汆水。
2. 全部材料用 1.2 公升水煮個半小時，調味即成。

食療功效

健脾補腎、養心安神。

飲食宜忌

本品清香，老少皆宜。適合手術前或手術後出現貧血、心悸怔忡、失眠健忘、虛汗頻出、神經衰弱者服用。但便秘及外感未清者不宜。

認識主料

蓮子

有鮮蓮子、乾蓮子，還有石蓮子。鮮蓮子最清香，但不是長年有供應；一般用來煲湯的多用開邊有衣蓮子，健脾補血功效佳，煲糖水則會用白蓮子。

蓮子茯神桂圓湯

Pork rib soup with lotus seed, Fu Shen and dried longans

薏米雙豆飲

Soy bean milk with mung bean and Job's tear puree

—— 手術前後提高抵抗力果汁・茶飲 ——

材料（1人量）

綠豆	30 克
薏米	30 克
無糖黑豆漿	250 毫升

做法

1. 綠豆、薏米用清水浸過面半天，放入攪拌機中，加入黑豆漿攪成糊。
2. 將綠豆薏米糊煮 10 分鐘可飲用。

薏米

能清熱利濕，除風濕，利小便，益肺排膿，健脾胃，強筋骨。能治風濕身痛、腳氣、筋急拘攣、水腫、肺萎、咳吐膿血等。但孕婦、津枯便秘及小便多者不宜。

手術前後提高抵抗力 果汁・茶飲

食療功效

清熱解毒、消脂降壓。

綠豆

能清熱解毒，消暑除煩渴。綠豆的藥理作用為降血脂、降膽固醇、抗過敏、抗菌、抗腫瘤、增強食欲、保肝護腎。但體質偏寒及痛風患者不宜。

飲食宜忌

本品能補充體力、增強體質，更有助降低血糖及膽固醇。適合煩躁口渴、二便不暢、皮膚瘙癢者手術前及手術後飲用。但脾胃虛寒者不宜。

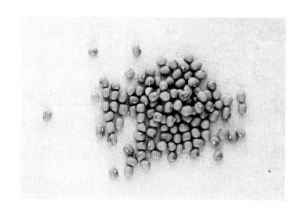

蘋果蔬菜汁

Apple and veggie juice

材料（1人量）

蘋果	1個
紅蘿蔔	1個
薯仔	1個

做法

1. 紅蘿蔔、蘋果、薯仔洗淨後抹乾，去皮，切塊。
2. 將切好的材料放入榨汁機內，榨成汁即可空腹飲用。

食療功效

排清毒素、防癌抗癌。

飲食宜忌

本品宜即榨即飲，切勿存放。宜每天飲用，連服一至三月，對防治癌症、保護肝臟，及手術前和手術後增強體力很有神益。但腎病患者不宜過量飲用，以免高鉀飲食加重腎臟負擔。

小貼士

最好選用有機蔬果，去皮後必須清洗乾淨，抹乾才榨汁。任何細菌、蟲卵都會影響病者健康。發芽薯仔含有龍葵鹼毒素，不宜用來榨汁。

手術前後提高抵抗力 果汁・茶飲

───── 認識主料 ─────

蘋果

蘋果功能補氣健脾、生津、止瀉。對消化不良、腹瀉等症很有幫助，蘋果同時可作為各種疾病治療期及癒後的營養補充劑。但胃腸炎者過量食用易導致腹瀉、口中泛酸。

薯仔

薯仔功能健脾、補氣、解毒。對胃熱痛，口泛酸水，食欲不振者很有幫助。但發芽及皮帶青色的薯仔含有龍葵鹼毒素，不宜食用。

材料（1人量）

馬蹄	6 粒
蓮藕	150 克
雪梨	1 個

做法

1. 馬蹄和蓮藕去皮洗淨，切碎。雪梨去皮，洗淨，切開，去核後才切碎。
2. 將全部材料榨成汁，放入煲內加熱 5 分鐘可飲用。

手術前後提高抵抗力 果汁・茶飲

───│食療功效│───

養陰潤燥、清心除煩。

───│飲食宜忌│───

本品甘甜，對手術前及手術後津液虧損、口燥咽乾、五心煩熱、大便秘結，尤其是對肺臟手術患者很有裨益。但脾胃虛寒、大便溏薄者不宜飲用。

Tips
小貼士

馬蹄、蓮藕均屬水生植物，容易受細菌感染及蟲害入侵，故不宜生食。蓮藕忌鐵器，故榨汁後最好用瓦煲或康寧煲烹煮。

馬蹄蓮藕甘露飲

Water chestnut, lotus root and pear juice

馬蹄

具有清肺熱、生津潤肺、化痰利腸、通淋利尿、清音明目等作用。對熱病消渴、目赤、咽喉腫痛、小便赤熱短少、燥熱咳嗽等有療效。但脾胃虛寒泄瀉、肺寒咳嗽、小兒遺尿者慎服。

奇異果紅棗茶

Red date and kiwi tea

▌材 料（1 人 量）

奇異果	2 個
紅棗	30 克
紅茶	3 克

▌做 法

1. 奇異果洗淨去皮，切碎；紅棗沖洗淨，去核。
2. 將奇異果、紅棗用 3 碗水煮 15 分鐘，加入紅茶，熄火焗 5 分鐘可飲用。

食療功效

生津利尿、健腦抗癌。

飲食宜忌

本品空腹飲用除了養顏之外，還可以幫助腸胃蠕動，清除宿便，適合癌症、心血管病、血壓高及食欲不振人士手術前及手術後飲用。但腎功能衰竭及有些人對奇異果過敏，尤其是嬰幼兒就不宜服。

認識主料

奇異果

可養顏美容、助消化、抗老、增強免疫力、降低膽固醇；豐富的膳食纖維能促進腸道蠕動，改善便秘。其所含的血清素具有穩定情緒、鎮靜心情的作用，有助於腦部活動。金黃色果肉的金奇異果味道較為清甜。

果仁水果奶

Walnut and cashew milk with apple puree

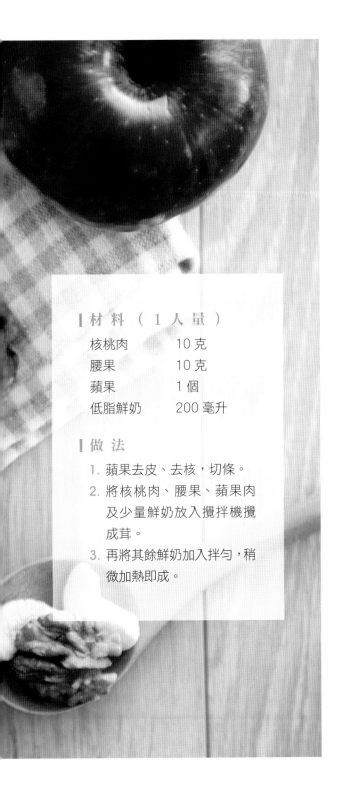

材料（1人量）

核桃肉	10 克
腰果	10 克
蘋果	1 個
低脂鮮奶	200 毫升

做法

1. 蘋果去皮、去核，切條。
2. 將核桃肉、腰果、蘋果肉及少量鮮奶放入攪拌機攪成茸。
3. 再將其餘鮮奶加入拌勻，稍微加熱即成。

食療功效

健脾補腎、強心養神。

飲食宜忌

本品營養豐富，能增強體質，延緩衰老。適合糖尿病、心血管病、腦力衰退者手術前及手術後飲用。但腹瀉及對鮮奶過敏者不宜。

認識主料

核桃

含有豐富的脂肪和蛋白質，多吃核桃可增強大腦的活力，且有補腎固精功能。不要剝掉核桃肉那層薄薄的褐色衣，它含有豐富的營養素，宜一起食用。

材料（1人量）

香蕉	1根
豆漿（室溫）	200毫升

做法

1. 香蕉去皮，切塊；用少許豆漿一同放入攪拌器中略為打散。
2. 將打散的香蕉豆漿倒出，再加入其餘豆漿拌勻即可供飲。

香蕉豆漿

Banana soymilk

食療功效

補虛潤燥、疏通血脈。

飲食宜忌

本品清香美味，但豆漿宜放至室溫才製作，不宜冷凍飲用。對手術前或手術後熱病煩渴、腸燥便秘、心血管病、皮膚乾燥、肺熱咳嗽者有益。但脾胃虛寒、夜尿多及急、慢性腎炎患者不宜。

認識主料

香蕉

它是調理腸胃失調的食物，纖維柔軟而柔滑，對於長期病患者來說，香蕉是唯一可以進食的未煮熟食物，而不會有不良反應。香蕉還可以中和胃酸和減少疼痛。

豆漿

有補虛潤燥、清肺化痰的功效。它的鐵含量高於牛奶五倍，其營養成分雖好，但性質微寒而滑削，胃寒者最好加些薑汁同服。

檸蜜百香果飲

Honey lemon passionfruit tea

手術前後提高抵抗力果汁・茶飲

材料（1人量）

檸檬	2片
蜂蜜	適量
百香果	1個

做法

1. 整個鮮檸檬投入開水中浸片刻，沖洗去掉表皮的蠟及農藥。
2. 檸檬抹乾，切片，放入玻璃瓶內，加入蜂蜜浸過面，醃1晚。
3. 取2片檸檬及少許醃過的蜂蜜放入杯內，刮入一個百香果果肉入杯內，用溫水沖服即成。

食療功效

鎮靜神經、健胃消炎。

飲食宜忌

本品能增強免疫功能，調整血液循環，而且能舒緩抑鬱，對手術前或手術後精神緊張，食欲欠佳者都有幫助。但對百香果過敏者、腸胃不太好及胃酸過多者慎用。

認識主料

百香果

它是天然鎮靜劑，具有鬆弛神經的功效。百香果果肉內的黑色種子一般可以食用，但不易消化；因此胃潰瘍、胃炎患者最好不要吃種子。

紅棗糙米茶

Red date brown rice tea

材料（1人量）

紅棗	4 粒
糙米	2 湯匙

做法

1. 紅棗去核，切薄片；糙米用白鑊慢火炒 7 分鐘。
2. 將材料放入壺內，注入開水，焗15分鐘可飲用。

食療功效

健脾補血、益智安神。

飲食宜忌

本品清香，老少皆宜。適合高血脂、高膽固醇、貧血、肥胖、心血管病、腦力衰退、癌症患者手術前或手術後飲用。但手術後消化不好的人，只宜飲茶不宜吃糙米。

認識主料

糙米

營養較精白米高，含大量纖維素，能改善腸胃機能、淨化血液、預防腸癌、健腦益智、幫助新陳代謝等。糙米炒香後放玻璃瓶儲存，可減少蟲患，並可隨時取用。

四紅茶

Four-red tea

▎材料（1人量）

紅衣花生	30 克
紅豆	30 克
杞子	6 克
紅棗	4 粒

▎調味料

紅糖	1 湯匙

▎做法

1. 紅豆、紅衣花生、杞子分別浸洗；紅棗去核。
2. 全部材料用 5 碗水煮 1 小時，調入紅糖煮溶即成。

食療功效

補血養肝、健脾安神。

飲食宜忌

本品香甜美味，老少皆宜。適合血虛、面色無華、水腫、失眠，以及遷延性肝炎病人手術前及手術後服食。但糖尿病、跌打瘀腫人士不宜。

認識主料

紅衣花生

能抑制纖維蛋白的溶解，增加血小板的含量，改善血小板的質量，促進骨髓造血機能。所以對各種出血及出血引起的貧血、再生障礙性貧血等疾病有明顯效果。紅衣花生不宜血液黏稠度偏高的人食用，否則易引發血栓。跌打瘀腫及消化不良者不宜食。

紅豆

功能理氣活血、清熱解毒。對心胃氣痛、疝氣疼痛、血滯經閉有幫助。常吃紅豆有助淨化血液、改善疲勞。但有頻尿困擾的人在食用上要多節制。

醋溜魚塊

Poached fish fillet in vinegar sauce

材料（3～4 人量）

急凍去骨魚柳	150 克
紅、黃色甜椒	各 1/4 個
雲耳	3 克
芫茜	1 棵
薑	2 片

醃魚肉料

鹽	1/6 茶匙
胡椒粉、生粉及米酒	適量

調味料

米醋	2 湯匙
赤砂糖	1 湯匙
鹽	1/4 茶匙
生抽	2 茶匙
胡椒粉	少許
清水	30 毫升

做法

1. 魚柳解凍後切片，用醃料醃半小時；紅、黃甜椒洗淨，切塊；雲耳浸軟，去蒂；芫茜洗淨，切碎。
2. 燒熱水，將魚肉投入即熄火，浸片刻撈起。
3. 鑊中放少許油，爆香薑片，放入甜椒、雲耳炒香，潷酒加調味，待汁滾起，放入魚片及芫茜，炒勻即可上碟。

認識主料

雲耳

黑木耳、雲耳同屬木耳科，雲耳質地較黑木耳柔軟，口感非常好。含有豐富的膠質，對人體消化系統有良好的清潤作用，具有清毛塵、洗腸、潤肺、減少血液凝塊、緩和冠狀動脈粥狀硬化、降低血栓的作用。

───── 食療功效

補血健脾、滋補強身。

───── 飲食宜忌

本品酸甜醒胃,美味可口,老少皆宜。對高血壓、高膽固醇、心絞痛、癌病患者手術前或手術後食慾欠佳、神疲乏力、大便秘結者有幫助。但骨折、跌打損傷、服薄血藥人士不宜食雲耳。

Tips 小貼士

急凍魚肉可以用石斑塊、青衣魚柳、太平洋鰈魚柳,或去骨銀鱈魚扒。魚肉解凍後先稍醃,然後投入滾水中汆水,魚肉香滑不會韌。用醋溜的烹煮法,可增加鈣質的吸收。

茭白香芹炒雞柳

Stir-fried shredded chicken with
Chinese celery and water bamboo shoots

材料（3～4人量）

雞柳	2 條
茭白	200 克
香芹	30 克
甘筍	1 小段
薑茸	1 茶匙

醃肉料

鹽、胡椒粉、米酒、粟粉	各少許

調味料

鹽	半茶匙
蠔油	1 茶匙
米酒	半湯匙

做法

1. 雞柳洗淨，切粗絲，用醃料醃半小時；甘筍切花；茭白去外殼，洗淨切粗絲；香芹洗淨切段。
2. 燒熱少許油，爆香薑茸，加入雞柳炒至肉變白色盛起。
3. 鑊中有少許油，加入茭白絲、香芹、甘筍兜炒，雞柳回鑊，潲酒，加調味，炒至汁收乾即可上碟。

食療功效

清熱解毒、降壓除煩。

飲食宜忌

本菜清香味美，老少皆宜。適合高血壓、糖尿病、濕熱黃疸、目赤腫痛者手術前或手術後食用。但脾胃虛寒及尿路結石者不宜。

認識主料

茭白

嫩茭白味道鮮美，人體容易吸收；但茭白含有較多的草酸，不能與豆腐同食，以免造成草酸鈣形成結石。茭白出現黑點是染了菌，絕對不宜食用。

做法

1. 南瓜洗淨，去皮，切塊，放入蒸碟內墊底；
2. 冬菜洗淨榨乾，與壓碎了的豆腐、薑茸、葱花、醃肉料同放入絞碎肉中攪成泥狀，搓成肉丸，排放在南瓜面。
3. 用大火蒸約 20 分鐘即成，下已拌勻的調味料。

│材料（3～4人量）

南瓜	360克
冬菜	半湯匙
硬豆腐	半磚
薑茸	1茶匙
葱花	2茶匙
絞碎肉	250克

│醃肉料

蛋白	1個
鹽	半茶匙
生抽	半湯匙
粟粉	1茶匙

│調味料

蠔油	1茶匙
生抽	半湯匙

認識主料

南瓜

南瓜含豐富的果膠；圓身桔紅色的南瓜，瓜味較濃及味道較甜；糖尿病患者宜選長身、糖分較低的南瓜。

豆腐肉丸蒸南瓜

Steamed pumpkin with tofu pork balls

温中益氣、降低血糖。

本品鮮甜美味，老幼皆宜。適合糖尿病、心血管病、腸燥便秘者
手術前或手術後食用。但氣滯濕盛者不宜多食。

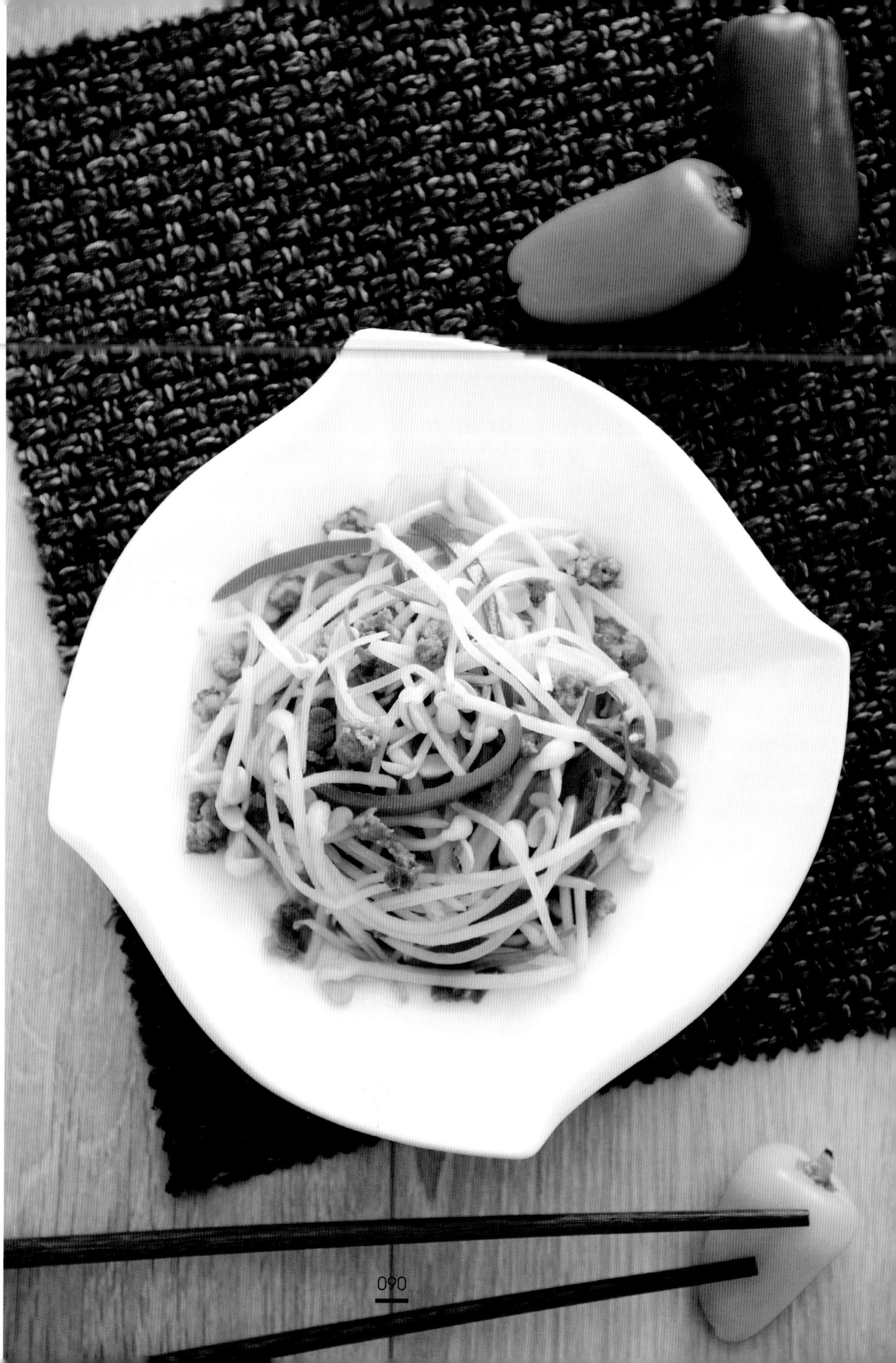

豆芽豬肉鬆

Stir-fried ground pork with soybean sprouts

▌材料（3～4人量）

大豆芽菜	250 克
絞碎豬肉	100 克
紅色甜椒絲	1 湯匙
葱絲	1 湯匙

▌醃肉料

| 生抽、胡椒粉、粟粉 | 各少許 |

▌調味料

鹽	1/4 茶匙
蠔油	1 茶匙
麻油	少許

▌做 法

1. 絞碎肉用醃料醃入味；大豆芽菜洗淨，去根，剁碎。
2. 大豆芽菜用白鑊炒至乾水，盛起。
3. 燒熱少許油，將絞碎肉炒香，再加入大豆芽菜、甜椒絲及調味兜炒至肉熟，最後灑入葱絲，炒片刻即成。

食療功效

健脾補中、滋陰潤燥。

飲食宜忌

本菜屬家常小菜，能滋養陰液，老少皆宜。適合形體虛弱、胃納欠佳者手術前或手術後食用。對癌症電療、化療後咽乾口燥者也十分適合。痛風患者宜少食豆類食物。

認識主料

大豆芽菜

大豆芽菜即黃豆芽，自然培育的黃豆芽豆粒正常、色澤金黃、芽桿直而稍細，根鬚長，無爛根或爛尖現象；經化學肥浸泡的黃豆芽芽桿粗壯，根短或無根。

合掌瓜炒玉蘭片

Stir-fried kale stems with chayote

—— 手術前後營養小菜 ——

材料（3～4人量）

材料	分量
合掌瓜	1個
芥蘭度	100克
金針	3克
甘筍	1小段
薑茸	1茶匙
上湯	30毫升
米酒	1湯匙
生粉水	1湯匙

調味料

調味料	分量
鹽	半茶匙
糖	1/4茶匙

做法

1. 合掌瓜去皮、去核，切片；金針浸軟後打成結；芥蘭度洗淨，切片；甘筍切花。
2. 燒熱少許油，爆香薑茸，倒入芥蘭度、合掌瓜片、金針炒香，潛酒，加調味及上湯，兜炒片刻，調入生粉水炒勻即成。

食療功效

舒肝理氣、清熱消脂。

飲食宜忌

本菜清香可口，老少皆宜。適合肝氣不舒、膽結石、膽總管及肝管結石了術前或手術後氣虛血弱者食用。但胃寒口泛清涎者宜少食。

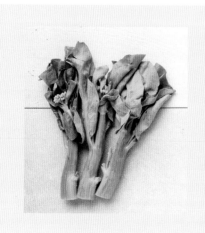

認識主料

芥蘭

芥蘭能解毒利咽、順氣化痰。含大量膳食纖維，有助排便；所含芥蘭素，可幫助記憶，但過食會耗人真氣，抑制性激素分泌；故孕婦不宜多食芥蘭，甲狀腺失調者亦不宜多食。

錦繡蛋絲

Shredded omelette with assorted veggies

材料（3~4人量）

雞蛋	3 個
鮮茴香莖	半個
香菇	1 朵
紅蘿蔔	1 小段
青瓜	1 小段

蛋料

鹽	1/4 茶匙
麻油	1 茶匙

調味料

鹽	半茶匙

做法

1. 雞蛋打散，加入鹽和麻油攪勻；茴香莖洗淨，切幼絲；香菇浸軟，去蒂後切幼絲；紅蘿蔔去皮洗淨，切絲；青瓜去皮洗淨，切絲。

2. 用平底鑊，燒熱油，先將雞蛋煎香盛起，切絲。

3. 鑊中留少許油，加入茴香絲、香菇絲、紅蘿蔔絲及青瓜絲炒香至汁收乾，最後加入蛋皮絲及調味，炒勻即可上碟。

認識主料

茴香莖

茴香能溫腎散寒、和胃理氣，常食有助促進消化，健脾暖胃，可緩解胃腸痙攣、減輕疼痛，亦適合白血球減少症患者。它含有的茴香素可對付癌細胞，尤其是胰臟癌。茴香更能穩定血糖，故對癌症、糖尿病、腎虛腰痛、脾胃虛寒、腸絞痛及痛經患者有益。但陰虛火旺者不宜食。

新鮮茴香莖在超市及售賣外國菜的菜檔有售。

手術前後營養小菜

食療功效

健腦益智、增強體質。

飲食宜忌

本品清香味美，老少皆宜。適合食欲不振、氣血虛弱、記憶力減退者手術前或手術後食用。一般人士皆可食。

五彩甜酸鮮淮山

Stir-fried fresh yam with five-colour veggies

材料（3~4人量）

鮮淮山	250 克
紫洋葱	30 克
鮮菠蘿	30 克
青、紅甜椒	各 10 克
薑茸	1 茶匙

▌醃淮山料

鹽	1/4 茶匙
生粉	1 湯匙

▌調味料

米醋	2 茶匙
赤砂糖	2 茶匙
番茄醬	1 湯匙
鹽	1/4 茶匙
生抽	半湯匙
粟粉	適量

▌做法

1. 鮮淮山去皮，洗淨後切件，用醃料醃片刻；紫洋葱去衣，切片；鮮菠蘿切小塊；青、紅甜椒切件。
2. 用平底鑊，燒熱油，將鮮淮山煎至金黃，盛起，其餘材料放入鑊中炒香，鮮淮山回鑊，加調味煮至汁濃稠即可上碟。

健脾益胃、幫助消化。

本品甜酸可口，非常醒胃。對手術前後食欲欠佳、消化力弱、體倦乏力、脾虛泄瀉等有益。但感冒、腸胃積滯者忌食。

淮山

能健胃厚腸、補肺益腎。對脾虛泄瀉、久痢、虛勞咳嗽、遺精帶下、小便頻數、消渴、子宮脱垂者有幫助。鮮淮山含有黏液蛋白，有助防治動脈粥樣硬化，故不要洗去黏液。

鮮淮山可當作蔬菜食用，有降血糖作用；以鐵棍淮山味道香糯美味，療效亦佳。但濕熱實邪及便秘者忌食。

手術前後營養小菜

097

素炒豆乾絲

Stir-fried dried tofu with assorted vegetables

▌材 料（3～4人量）

黑木耳	3 克
紅蘿蔔	半條
荷蘭豆	30 克
五香豆乾	3 塊
薑茸、蒜茸	各 1 茶匙
紹酒	2 茶匙
生粉水	1 湯匙

▌調 味 料

鹽	半茶匙
糖	半茶匙
蠔油	2 茶匙

▌做 法

1. 黑木耳浸軟，去蒂後切幼絲；紅蘿蔔去皮，切絲；荷蘭豆撕去老筋，洗淨切絲；豆乾沖洗後切絲。
2. 燒熱油，放入豆乾慢火煎至金黃，盛起；鑊中留少許油，爆香薑茸、蒜茸，放入黑木耳絲、紅蘿蔔絲及荷蘭豆絲炒至熟。
3. 加入豆乾絲，灒酒加調味，最後加入生粉水，兜炒至勻即成。

___食療功效___

平肝清熱、增進食欲。

___飲食宜忌___

此菜含豐富纖維，有助潤腸通便、修身減肥。適合肥胖、高血脂、高膽固醇人士手術前或手術後食用。但服薄血藥者不宜吃黑木耳，因黑木耳含抗凝血物質；因此，手術前2-3天及手術後一週都不宜用黑木耳，可用冬菇代替。

___認識主料___

黑木耳

黑木耳有「血液清道夫」之稱。有祛瘀、降膽固醇、滋補強壯、清肺益氣、補血活血、涼血止血、鎮靜止痛的功效。含維生素 K，能減少血液凝塊，預防血栓症發生。故出血性疾病患者不宜食。

黑木耳以朵面大而乾淨、光滑油潤、面呈黑色、底呈灰白、浸泡後浮起不黏手為佳。

材料（3～4人量）

小花菇	6-8朵
浸發海參	2條
薑絲	半湯匙
白菜菜膽	6棵
上湯	1碗
米酒	1湯匙
生粉水	1湯匙

調味料

鹽	半茶匙
糖	半茶匙
蠔油	2茶匙

紅燒花菇海參

Braised sea cucumbers with
shiitake mushrooms

▌做 法

1. 花菇浸軟，去蒂；浸發海參切塊後汆水。
2. 燒熱少許油，爆香薑絲，加入花菇、海參炒香，讚酒，加調味及上湯煮熟。
3. 菜膽洗淨，投入油鹽水中焯至碧綠剛熟，排碟邊。
4. 花菇、海參燜煮大約半小時，調入生粉水至汁濃稠，盛入碟中即成。

食療功效

補益肝腎、防癌抗癌。

飲食宜忌

此菜香滑美味，老少皆宜。適合高血壓、高膽固醇、肺結核、肝炎、神經炎、血友病、癌症等患者手術前或手術後食用。痰濕壅滯、便溏腹瀉及痛風患者忌食。

認識主料

海參

有補腎益精、養血潤燥、止血消炎的功效。對精血虧損、虛弱、消渴、腸燥便秘、皮下出血、血管硬化等均甚有益。海參不含膽固醇，更屬陰陽雙補之品，無論腎陰虛、腎陽虛體質皆可服。但痛風患者不宜。

海參有很多品種，其中以刺參最香滑，禿參適合作家常小菜，豬婆參處理費時，較適合酒樓食店應用。選購以直身、無破損、重手者為佳。

蒸釀節瓜環

手術前後營養小菜

▌材料（3~4人量）

節瓜	1 大條
絞碎豬肉	200 克
蝦米	1 湯匙
葱花	1 湯匙

▌醃肉料

蛋白	1 個
鹽、胡椒粉、粟粉	適量

▌芡料

生抽	2 茶匙
米酒	半湯匙
糖	1/4 茶匙
生粉水	1 湯匙

做法

1. 節瓜刮去外皮,切成2公分厚的圓環,挖去瓜籽,投入油鹽開水中燙半分鐘,撈起瀝乾。
2. 蝦米浸軟,剁碎後和葱花同加入絞碎肉中,加入醃肉料攪拌成膠。
3. 在節瓜內圈抹點粟粉,填入肉料,放蒸碟中,隔水蒸25分鐘,將汁倒出,勾芡即成。

食療功效

清熱利水、滋陰潤燥。

飲食宜忌

本菜清甜美味。適合精神困倦、食欲不振、喉乾咽燥及虛不受補者手術前或手術後食用。一般人群皆可食。

認識主料

節瓜

節瓜是冬瓜的變種,但少了冬瓜寒涼之氣,是有益正氣的瓜菜。對身體虛弱而有發熱現象,出現口乾口渴、煩躁、二便不暢者很有幫助。一般人士皆可食用。

手術前後營養小菜

手術甦醒後清流質飲品

金蕎麥杞子茶

Bitter buckwheat tea with Goji berries

材料（1人量）

金蕎麥	2 湯匙
杞子	1 茶匙

做法

1. 金蕎麥、杞子同放入茶包袋，將茶包放入壺內，用開水沖洗一次。
2. 再注入開水，焗 10 分鐘即可飲用。

食療功效

清熱解毒、健脾利濕。

飲食宜忌

本茶清香味美，對三高症人士及任何患者手術後煩躁失眠、排尿困難、面色蒼白等均有益。但體質偏寒及夜尿多者慎服。

認識主料

金蕎麥

分甜蕎麥和苦蕎麥。金蕎麥即苦蕎麥，金蕎麥含有生物類黃酮蘆丁。蘆丁有軟化血管、改善微循環、降低血糖、降低血脂、降低膽固醇等功效。其療效較甜蕎麥更為優勝。

小麥黑豆安神茶

Wheat groats and black bean tea

■ 材料（1人量）

小麥米	30克
青仁黑豆	30克
茯神	10克
圓肉	8粒

■ 做法

1. 小麥米、黑豆、茯神分別浸洗，圓肉沖洗。
2. 用5碗水煮1小時即成。

食療功效

養心安神、祛風通絡。

飲食宜忌

本品清香，對任何手術後心煩不寧、多夢易醒、多汗、血虛面色蒼白、關節疼痛者很有幫助。但痛風人士不宜吃豆。

109

黑豆

有活血、利水、祛風、清熱解毒、滋養健血、補虛烏髮的功能。黑豆同時有降低血中膽固醇的作用。因此,常食黑豆,能軟化血管,滋潤皮膚,延緩衰老。選購黑豆時,以黑皮青肉的為佳,青仁黑豆兼具滋補肝腎之功。但痛風症患者不宜食用。

認識主料

茯神

是茯苓菌中間部位,寧心安神功效佳,更有利尿作用,能促進麻醉藥的排出。

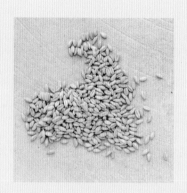

小麥米

能養心益腎、清熱除煩。小麥米可以降低血液循環中的雌激素的含量,從而達到防治乳腺癌的目的,且能緩解更年期綜合症。一般人群可食,但胃寒者宜少食。

蘿蔔陳皮水

Radish and dried tangerine peel tea

▌材 料（１人量）

白蘿蔔	1 條
陳皮	? 個

▌做 法

1. 白蘿蔔洗淨，去皮，切塊；陳皮浸軟洗淨。
2. 白蘿蔔、陳皮用 5 碗水煮半小時成 2-3 碗，即可供飲。

食療功效

益脾和胃、通利二便。

飲食宜忌

本品大口大口飲能促進腸胃蠕動，幫助排便；小口小口飲能促進膀胱活動，加快排走手術後滯留在體內的麻醉藥，使體內毒素及早排出體外，同時還有清痰作用。適合任何手術後病患飲用。

認識主料

蘿蔔

清熱生津，化痰止咳，通利二便。選購蘿蔔時，以表皮光滑、根部大而圓厚、根鬚少為優，同時握在手裏沉甸甸的，這可避免買到空心蘿蔔。

陳皮

有行氣健脾、降逆止嘔、調中開胃、燥濕化痰的功效，適用於脾胃氣滯、脘腹脹滿、噁心嘔吐、咳嗽痰多等症狀。陳皮與維生素 C、維生素 K 食物同用，能增強抗炎作用。但有胃火的人不宜多食用。

手術甦醒後清流質飲品

檸檬醒脾飲

Brown sugar lemonade

材料（1人量）

鮮檸檬　　　　半個

調味料

紅糖　　　　　2茶匙

做法

1. 鮮檸檬投入開水中浸片刻，再沖洗（洗去蠟和農藥）。
2. 取半個檸檬，切片，放入壺內，加入紅糖及開水，焗5分鐘可供飲用。

手術甦醒後清流質飲品

食療功效

開胃醒脾、利尿生津。

飲食宜忌

本品一般人均可飲用，尤其對手術後口乾煩躁、精神未恢復、神疲乏力者有幫助。任何手術後都可以飲用，但胃潰瘍、胃酸分泌過多者不宜。

認識主料

檸檬

檸檬的清香味有助醒脾，進入體內有助身體淨化，使排尿速度增加。因此，麻醉藥毒素能以更快的速度釋放排掉。

宜挑選兩頭較圓的檸檬，除了沒那麼酸外，果汁會多一點；兩頭比較尖的，會比較酸，而且果汁也會少一點。

金針紅棗水

Red date and day lily tea

▌材料（1人量）

金針　　　　5克
紅棗　　　　4粒

▌調味料

紅糖　　　　1茶匙

▌做法

1. 金針浸洗；紅棗沖洗後去核，切片。

2. 將金針、紅棗用2碗半水煮10分鐘，加入紅糖煮溶即可供飲。

紅棗

能補中益氣、養血安神、調和藥性，且能延緩衰老、抗疲勞、保護肝臟、抗腫瘤、增強機體免疫力，可治療貧血、虛寒、腸胃病等。但過量食用會容易脹氣，使人肥胖，減重者忌食。

金針

又名「忘憂草」，有安神作用。其鐵質含量為菠菜的 20 倍，是補血的最佳食材之一，亦可治大便下血及其他出血症。選購時以色澤鮮明帶黃、無受潮、無酸味者為佳。

食療功效

補血活血、安神止血

飲食宜忌

本品香甜可口，對手術後虛弱、容易疲倦、面色蒼白、血壓低、小便不利、記憶力衰退者很有益。任何手術人士可服。

雪耳蛋花湯

Egg drop soup with white fungus

材料（2人量）

雪耳	6克
鮮百合	20克
雞蛋	1個
杞子	1茶匙
上湯	2碗

調味料

海鹽	半茶匙

做法

1. 雪耳浸洗淨，去蒂後剁碎；鮮百合剝開，洗淨；雞蛋打散。

2. 燒熱上湯，加入雪耳、鮮百合和杞子煮30分鐘，最後加入調味及蛋液，滾起即成。

手術後半流質飲食

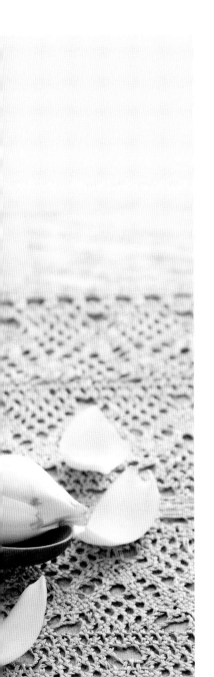

食療功效

滋陰潤肺、清心安神。

飲食宜忌

本品香滑味美，能補肝虛又不會引動肝火。對手術後睡眠欠佳、精神不振、陰虛火旺、肺熱咳嗽、痰中帶血者有益。一般手術患者皆可食用，但服薄血藥者忌食。

認識主料

雪耳

雪耳不但營養豐富，更含植物性的化學成分，當中的蛋白質就含有 17 種氨基酸，有助促進新陳代謝。選購雪耳以乾燥、色澤潔白、肉厚、朵整、無刺激性氣味為佳，過白或過黃，並附有刺鼻氣味的都不適宜食用。

鮮百合

百合能清心除煩、寧心安神。鮮百合含黏液質，具有潤燥清熱作用，對肺燥或肺熱咳嗽，化療及放射性治療後白細胞減少有治療作用。但風寒咳嗽、虛寒出血症者忌食。

紫菜粟米豆腐羹

Tofu thick soup with laver and sweet corn

材料（2人量）

紫菜	1 小撮
急凍粟米粒	1 湯匙
青豆粒	2 茶匙
嫩豆腐	1 盒
上湯	2 碗
生粉水	2 湯匙
蛋白	1 個

調味料

鹽	半茶匙

做法

1. 紫菜沖洗；急凍粟米粒、青豆粒解凍；豆腐沖洗後切粒。
2. 燒熱上湯，加入豆腐、粟米粒、青豆粒及紫菜煮 15 分鐘，加調味及生粉水，邊煮邊攪，最後加入蛋白，熄火焗片刻即成。

食療功效

清理腸胃、化痰散結。

飲食宜忌

本品香滑，適合頸部切面較深的手術、胃腸及腹腔手術後食欲不振、需要排膿引流者食用。一般手術患者可食，但脾胃虛寒者少食。

認識主料

紫菜

紫菜性寒，能軟堅散結，清熱化痰，利尿。常用於甲狀腺腫、水腫、慢性支氣管炎、咳嗽、腳氣、高血壓、高膽固醇等症。但脾胃虛寒者及消化力弱者忌食。紫菜以深紫色、雜質少、乾燥無受潮者為佳。

瑤柱陳皮粥糊

Blended congee with dried scallops
and dried tangerine peel

材料（1~2人量）

蒸煮好的瑤柱	3 粒
陳皮	1 小塊
白米	60 克

醃米料

油、鹽	各少許

做法

1. 陳皮浸軟，切絲；瑤柱拆絲；白米洗淨，用油、鹽略醃。

2. 用 6 碗水將全部材料放入電飯煲中煮成稀粥，用攪拌器將粥攪拌成糊狀即可供食。

食療功效

滋陰補腎、補中益氣。

飲食宜忌

本品鮮甜，易於吸收。對手術後體力未恢復、食欲不振、心煩口渴、失眠多夢者有益。一般人士可食用，但痛風症患者不宜食用瑤柱。

Tips 小貼士

白米洗淨後用少許油、鹽略醃才煮，粥煲好後會更為香滑。瑤柱買回來後最好先浸軟，連浸水隔水蒸1小時，然後分包存放冰格，可隨時取用，蒸好的瑤柱絲會香滑無渣。

認識主料

白米

白米含有醣類、維生素B群、維生素E、鈣、磷、鉀等營養素。加工過於精細的白米，會損失很多胚乳與糊粉層的營養成分，營養價值較低，最好與糙米等搭配食用，才能兼顧營養。

菠菜鱸魚濃湯

Perch and spinach thick soup

▌材 料（２人量）

菠菜	50 克
海鱸魚	100 克
淡奶	100 毫升
麵粉水	1 湯匙

▌調 味 料

油、鹽	各少許

▌做 法

1. 菠菜洗淨，切碎；海鱸魚隔水蒸熟，拆肉。

2. 用約 3 碗水加入菠菜及鮮魚肉煮 7 分鐘，加入淡奶、麵粉水及調味，煮至濃稠，熄火，用攪拌器攪成糊狀即成。

食療功效

養陰補血、加快康復。

飲食宜忌

本品備有豐富，對小孩或．对腸道

手術後大便不暢、排便疼痛出血、
貧血頭暈者均有益。但服薄血藥者
忌食菠菜，以免引起出血症狀。

認識主料

鱸魚

含豐富蛋白質及各種微量元素，具
補肝腎、益脾胃、化痰止咳等功
效，對孕婦胎動不安、產後乳汁
少均有幫助。海鱸魚療效較淡水
鱸魚好。

海鱸魚含豐富的蛋白質，魚皮又
含豐富膠質，能補充營養，調理
傷口，保健身體，對手術後深層
傷口癒合很有幫助。其他身體的
創傷，如燙傷、挫傷、撕裂傷、
撞傷和擦傷等，都可多吃鱸魚湯
以加速修復和復原。但有皮膚病
患者不宜食。

菠菜

菠菜含豐富葉酸、胡蘿蔔素及多種維生素和
礦物質。有補血止血、利五臟、通血脈、止
渴潤腸等功效。對病後或手術後貧血、肝虛
目昏，或夜盲症者很有幫助。但服用薄血藥
者忌大量食用。

Millet congee with lotus root starch

藕粉小米粥

材料（1～2人量）

藕粉	2 湯匙
杞子	10 粒
小米	60 克
白米	10 克

調味料

原味砂糖	適量

做法

1. 小米沖洗後用水浸半小時；白米洗淨；藕粉用冷水開好；杞子浸洗。

2. 用 6 碗水將小米、白米連浸的水煮 1 小時，加入杞子、藕粉水及砂糖調勻至糖溶即成。

清熱涼血、健脾益胃。

本品對需要半流質飲食者最為適合，不但能減少手術創口疼痛及出血問題，更有助減少排便疼痛和出血，助腸道恢復，對腸道手術如內外痔、肛瘻、肛周膿腫等手術者均有裨益。任何手術患者可食。

藕粉

藕粉有益胃健脾、養血補益、止瀉等功能。對食欲不振、脾虛泄瀉、吐血、高血壓、肝病、缺鐵性貧血、營養不良者均有益。但藕粉多數味甜，故肥胖、糖尿病患者宜少食。

藕粉含有大量鐵質和還原糖等成分，與空氣接觸後極易氧化變微紅色，但如果呈玫瑰紅色，可能加了色素染色而成。藕粉存放久了會由微紅變為紅褐色，只要不受潮霉變，可以食用。

小米

是最能健脾養胃的滋補米，含維生素 B_1、B_{12} 等，具有防止消化不良和口角生瘡的功效；同時可防止反胃、止嘔。還有滋陰養血的功能，適合老人、病人和婦女。但氣滯及體質偏寒者忌食。

勝瓜草菇魚湯米線

Rice noodles in fish broth with
angled luffah and straw mushrooms

材料（2人量）

絲瓜	半條
草菇	4 粒
杞子	4 克
鮮米線	200 克
鮮魚骨湯	3 碗

調味料

海鹽	半茶匙

做法

1. 絲瓜削皮，切件；草菇沖洗，剖開對半；杞子浸洗。

2. 將魚湯煮滾，加入絲瓜、草菇、杞子滾 10 分鐘，加入鮮米線及調味，滾起即成。

Tips 小貼士

有些淡水魚檔有新鮮鯇魚骨或鯪魚骨出售，買些來熬湯，鮮味又富營養。

食療功效

清熱化痰、通利腸胃。

飲食宜忌

本品清甜味美，老少皆宜。對手術後食欲不振、煩熱口乾、喉有濃痰、二便不暢者有益。但脾胃虛寒者不宜。

認識主料

絲瓜

能清熱化痰、涼血、解毒。絲瓜中含有干擾素的誘生劑，能刺激人體產生干擾素，對抗病毒、抗癌。老身絲瓜能通經絡、利血脈，治療筋骨酸痛。但絲瓜寒涼，多食易引起滑腸泄瀉。絲瓜含有植物黏液甚多，其種子有峻瀉作用，如貪爽口味鮮將絲瓜煮至半生熟食用，易引起肚痛腹瀉，故絲瓜必須煮熟才食。

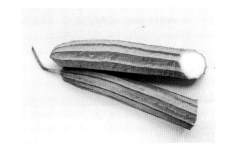

Steamed rice with pumpkin and raisins

南瓜葡萄乾飯

材 料（2 人 量）

南瓜	150 克
白米	1 杯
葡萄乾	2 湯匙

醃 米 料

油、鹽	各少許

做 法

1. 南瓜去皮，去籽，切粗粒；白米洗淨，用少許油、鹽略醃。
2. 南瓜與白米加1杯水煮成飯，拌入葡萄乾即成。

食療功效

補中益氣、排毒養顏。

飲食宜忌

本品香甜美味，老少皆宜。對手術後氣血不足、貧血、面色蒼白、咽乾煩渴、二便不利、風濕痹痛者有益。一般手術患者可食，濕重者不宜食。

認識主料

葡萄乾

葡萄乾補益氣血、滋陰生津、強筋健骨、通利小便。對氣血虛弱、肺虛久咳、肝腎陰虛、心悸盜汗、腰腿酸痛、小便不利等有益。葡萄還中含有一些抗癌物質，防止癌細胞擴散。但糖尿病患者忌食。南瓜含有豐富果膠，果膠能黏結體內細菌毒性和其他有害物質，如鉛、汞、鎘和放射性元素等，並能促進胃腸道潰瘍癒合，故很適合手術後食用。南瓜飯中加入葡萄乾，除可益補外，更能增加甜味及口感，但宜煮好飯才拌勻食，否則煮太久葡萄乾會變酸。

雞茸粟粒麥皮

Oatmeal with minced chicken and sweet corn kernels

材料（1～2人量）

新鮮雞柳	50 克
急凍粟米粒	2 湯匙
麥皮	20 克

醃肉料

蛋白	1 個
海鹽、粟粉	各少許

做法

1. 鮮雞肉剁碎，用醃肉料醃入味；粟米粒解凍。
2. 燒熱 2 碗水，加入粟米粒、麥皮，煮 5 分鐘，最後加入醃好的雞肉碎，待雞肉變白色即可熄火，焗片刻可食。

認識主料

雞柳

脂肪含量較少，蛋白質含量高，肉質細嫩。如果用蛋白等調味醃過，加上盡量縮短烹調時間，雞肉會更加嫩滑美味。

補中益氣、增強體力。

飲食宜忌

本品營養豐富，不肥膩又易消化，對手術後可以進食澱粉質食物的患者很有益。一般手術患者都可以食用，但腎臟手術病人及尿酸過高者宜少食麥皮。

手術後澱粉質主食

瑤柱蛋白薑茸炒飯

Fried rice with dried scallops, egg white and grated ginger

材料（2人量）

蒸煮好的瑤柱	3-4 粒
蛋白	2 個
薑茸	1 湯匙
芥蘭度	3 條
冷飯	2 碗
米酒	少許

調味料

海鹽	1 茶匙

做法

1. 蒸煮好的瑤柱拆絲；蛋白打勻；芥蘭度洗淨，切粒。
2. 燒熱油，爆香薑茸，加入芥蘭粒、瑤柱絲及冷飯炒香，灒酒加調味及蛋白液，兜炒至香味溢出、夠熱即成。

食療功效

滋陰潤燥、健脾暖胃。

飲食宜忌

本品鮮味可口，老少皆宜。對手術後體力未恢復、胃口未開、神疲乏力、心煩口渴者很有益。任何手術者皆宜食用。

小貼士

炒飯不黏鑊的秘訣是，先將鑊燒至很熱，倒一杓冷油入鑊再迅速將油倒出，令油溫不致過熱，然後加飯及餸料來炒，炒好的飯不會太油膩，而且不會黏鑊。

認識主料

瑤柱

有消脂降壓、消食、抑制腫瘤、養陰補虛的功效。對胃口欠佳、營養性水腫、氣血虛弱或病後、手術後體力未恢復需要調補者很適合。對失眠多夢、夜尿頻多等陰虛症亦有幫助。但痛風患者不宜食。

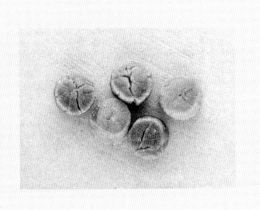

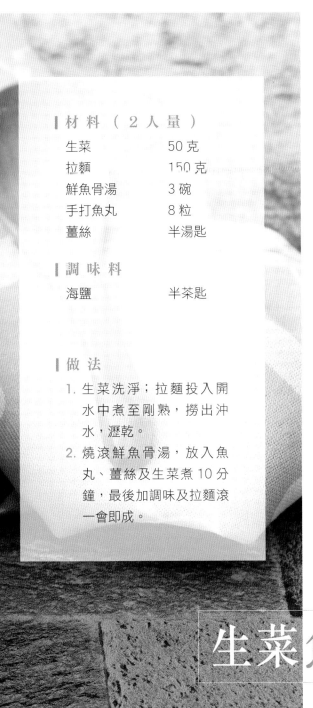

材料（2人量）

生菜	50克
拉麵	150克
鮮魚骨湯	3碗
手打魚丸	8粒
薑絲	半湯匙

調味料

海鹽	半茶匙

做法

1. 生菜洗淨；拉麵投入開水中煮至剛熟，撈出沖水，瀝乾。
2. 燒滾鮮魚骨湯，放入魚丸、薑絲及生菜煮10分鐘，最後加調味及拉麵滾一會即成。

食療功效

養血止血、清熱除煩。

飲食宜忌

本品鮮味可口，老少皆宜。對手術後腸胃燥熱、心煩口渴、貧血頭痛、大便澀滯不通者有益。一般人士可食用。

認識主料

麵食種類繁多，最好選購不經油炸，無過分加工，簡單以水及澱粉製成那些麵食較為健康。

生菜魚丸麵

Fish ball ramen with lettuce

圓肉紫米粥

Black glutinous rice congee with dried longans

—
手術後澱粉質主食
—

▌材 料（2人量）

圓肉	10 粒
紫米	100 克

▌做 法

1. 圓肉沖洗；紫米略為沖洗，用清水浸。
2. 圓肉、紫米連浸水放入煲中煮約1小時即可供食。

食療功效

補益心脾、養血安神。

飲食宜忌

本品香滑軟糯可口，對手術後精神不振、口涸咽乾、神疲乏力、貧血、流虛汗、心悸失眠者有益。任何手術患者可食。

認識主料

紫米

被稱為「補血米」，特別適合失血、體虛乏力、心悸氣短、流虛汗者食用。糯性紫米顆粒大而飽滿、黏性強，軟糯可口。紫米營養易溶於水，宜用冷水輕輕沖洗，不用揉搓，以免流失營養。

香菇滑雞粥

Chicken congee with shiitake mushrooms

手術後澱粉質主食

材料（2人量）

小香菇	4 朵
有機雞腿	1 隻
白米	60 克
薑絲、葱絲	各半湯匙

醃雞料

海鹽	半茶匙
胡椒粉	少許
粟粉	1 茶匙

做法

1. 香菇浸軟，去蒂切絲；雞腿起肉，切丁，用醃料醃入味；白米洗後用少許油、鹽略醃。
2. 將香菇絲、薑絲、白米及雞腿骨加水煮成濃稠適度的粥，將雞腿骨撈走，最後加入雞肉丁，煮5-6分鐘，灑入葱絲即成。

認識主料

冬菇

冬菇能益胃和中、化痰理氣。對治療食慾不振、身體虛弱、糖尿病、肺結核、傳染性肝炎、神經炎、小便失禁、大便秘結及癌腫患者有幫助。脾胃寒濕氣滯及痛風患者忌食。

食療功效

健腦益智、增強體力。

飲食宜忌

本品香滑可口，對手術後記憶力減退、失眠多夢、頭暈目眩、神疲乏力者有益。一般手術患者可食，但痛風患者不宜食菇類食物。

小貼士

有機雞肉特別鮮美，又少含激素。現時很多活雞檔都有新鮮有機雞，分不同部位斬件出售。用雞腿骨熬湯作粥底，更加鮮味。

手術後澱粉質主菜

Stir-fried spaghetti with
Iberian pork and beetroot

紅菜頭黑豚肉意粉

材 料 （ 2 人 量 ）	
紅菜頭	100 克
西班牙黑豚肉	100 克
珍珠筍	4 條
青椰菜	30 克
意大利粉	150 克

醃 肉 料	
生抽	1 茶匙
米酒	1 茶匙
粟粉	適量

調 味 料	
鹽	半茶匙
黑椒粉	少許

147

做法

1. 紅菜頭去皮，洗淨，切絲；黑豚肉切薄片，用調味略醃；珍珠筍洗淨，剖開對半；青椰菜洗淨，切絲。

2. 意大利粉投入放了少許鹽的開水中煮至夠熟，撈出沖水，瀝乾。

3. 鑊中放少許油，放入紅菜頭絲、意粉，加入黑椒粉及鹽炒至意粉變紅色，盛起。

4. 鑊中留少許油，放入黑豚肉炒香，再加入青椰菜絲、珍珠筍，兜炒片刻，盛起放在意粉旁即成。

食療功效

補血養顏、滋補強壯。

飲食宜忌

本品色香味全，營養豐富又美味，老少皆宜。對手術後或癌症患者出現氣血虧虛、面色蒼白、精神不振、食慾不佳、血壓不穩定者有益。但服薄血丸者忌用紅菜頭。

認識主料

紅菜頭

紅菜頭是護肝食物，它具有抑制血中脂肪、協助肝臟細胞再生與解毒的功能，所含葉酸豐富，補血功能很強。紅菜頭葉、莖的營養價值較紅菜頭更高，而且葉子可用來生吃，其食用纖維有助清腸排便。它更是防癌、抗衰老、抗自由基的最佳食品。一般人士可服。

紅菜頭與意粉同煮，意粉即染成桃紅色，美麗與營養兼備。

黑豚肉

黑豚肉味道鮮美、肉質細嫩，易消化吸收，屬低脂肪、低膽固醇肉類。具滋補養顏、壯陽等功效。對營養不良、貧血、失眠、胃病、高血壓、冠心病等有較明顯的食療作用，此外還有鋅、硒等具有抗癌作用的元素，是一種理想的天然滋補佳品。一般人士可服。

核桃雪耳燉海參

Double-steamed sea cucumber soup with walnuts and white fungus

材料（2人量）

核桃肉	20克
雪耳	9克
南棗	4粒
西施骨	150克
浸發海參	2條
生薑	3片

調味料

海鹽	1/4茶匙

做法

1. 核桃肉、雪耳浸洗，雪耳去蒂；南棗浸洗；西施骨、浸發海參汆水。
2. 全部材料放入燉盅內，注入3碗開水，燉3小時，調味即成。

認識主料

海參

海參屬陰陽雙補之品，無論腎陽虛、腎陰虛者均適合食用。海參含多醣物質，能抗放射損傷，促進造血功能、降血脂和有抗癌功效，對電療後虛不受補、氣血兩虛者甚為有益。

南棗

南棗有養脾、平胃氣、潤心肺、止咳嗽、補五臟、治虛損等功效，能養陰補血而不易上火。但濕盛及脘腹脹滿者忌食。

食療功效

健脾固腎、補血潤燥。

飲食宜忌

本品滋補美味，老少皆宜。對手術恢復期出現的氣血虧損、神疲乏力、失眠、記憶力衰退、小便頻數者有益。任何手術患者可食，但痛風者不宜食海參。

手術恢復期 滋補強身湯水

紅菜頭番茄薯仔瘦肉湯

Lean pork soup with beetroot, tomatoes and potatoes

▌材料（2人量）

紅菜頭	100 克
番茄	50 克
薯仔	50 克
瘦肉	150 克
生薑	2 片

▌調味料

海鹽	1/4 茶匙

▌做法

1. 紅菜頭、番茄、薯仔去皮，切塊；瘦肉切片，汆水。
2. 全部材料用6碗水煮1小時，調味即成。

食療功效

健脾和胃、消脂降壓。

飲食宜忌

本品甘甜，老少可服。對手術恢復期出現氣血虛弱、面色蒼白、血壓不穩、食欲不振、腸胃蠕動慢者有益。腎病、糖尿病、低血壓患者慎服。

紅菜頭

紅菜頭的部分營養在烹調過程中易於流失，而紅菜頭的枝葉含豐富的纖維素和各種營養素，故可將枝葉洗淨切絲或切粒製成沙律，這樣才不會浪費。

猴頭菇燉竹絲雞

Double-steamed silkie chicken soup with monkey-head mushrooms

材料（2 人量）

猴頭菇	2 朵
紅棗	6 粒
瑤柱	2 粒
有機竹絲雞	半隻
生薑	2 片

調味料

海鹽	1/4 茶匙

做法

1. 猴頭菇浸洗；紅棗去核；瑤柱浸軟；竹絲雞劏後洗淨，去皮，斬件後汆水。
2. 全部材料放入燉盅內，注入 3 碗開水，隔水燉 2 小時，調味即成。

食療功效

健脾養血、滋補肝腎。

飲食宜忌

本品清香味美，老少皆宜。對消化系統手術後體虛血虧，神疲乏力，或其他癌腫電療期間食少神疲、口乾渴飲、潮熱骨蒸均有益。但癌症屬濕熱毒盛者不宜飲用。

認識主料

竹絲雞

竹絲雞又稱烏雞，有補益肝腎、養陰退熱的功效。其肉味鮮美，既可作珍饈美饌，又被視為婦科聖藥，對婦女白帶症、不育症、月經不調、產後虛弱均有良效。亦是手術後滋補佳品。但外感發熱者不宜。

竹絲雞的營養價值遠高於一般雞隻，肉質細嫩。宜買無激素飼養、外國進口、有認證的急凍有機竹絲雞，在大型有信譽的急凍肉食超市有售。

猴頭菇

它能降低血液中膽固醇和甘油三酯含量，調節血脂，利於血液循環，防治心血管病，並能抑制癌細胞生長，防治癌症，尤其是消化道癌症。一般人士可食，但對菇類敏感者宜少食。

合掌瓜豆腐石崇魚湯

Scorpionfish soup with chayote and tofu

材料（2人量）	
合掌瓜	1 個
硬豆腐	1 磚
石崇魚	半斤
杞子	3 克
生薑	2 片

調味料	
海鹽	1/4 茶匙

—— 手術恢復期滋補強身湯水 ——

做 法

1. 合掌瓜去皮，切塊；豆腐沖洗切塊；石崇魚劏後洗淨，用少許油煎香。
2. 燒熱大滾水，放入全部材料煮約1小時，調味即成。

食療功效

理氣和中、滋補強身。

飲食宜忌

本品鮮甜美味，老少皆宜。對手術恢復期出現肝鬱氣滯、焦慮不安、脘腹脹痛、傷口疼痛者有益。任何手術患者均可服，尤其是剖腹產產婦，傷口痛、發炎，飲了石崇魚湯，康復得很快。一般人士可服。

認識主料

石崇魚

石崇魚有少許像石狗公，但形狀較醜陋。石崇魚功能健脾補腎，含豐富的蛋白質，對手術後深層傷口癒合很有幫助，但其刺有毒，被刺到會產生劇痛，故購買時最好叫魚檔代為處理。石崇魚配合營養成分較全面的合掌瓜同煮，清潤而不滋膩。一般人士可食。

杞子

能補肝明目，對腰膝酸軟、頭暈健忘、目眩、目昏多淚、消渴、遺精等症有益。常食有助提高身體免疫力，且能抑制癌細胞生長和突變，具有延緩衰老、防脂肪肝、調節血脂和血糖的功效。但感冒發燒、身體有炎症及腹瀉者忌用。

鮑魚杞子菊花湯

Abalone soup with Goji berries and chrysanthemums

▌材 料（2 人 量）

青邊鮑魚	1 隻
紅蘿蔔	1 條
杞子	5 克
菊花	6 克
生薑	3 片

▌調 味 料

海鹽	1/4 茶匙

▌做 法

1. 杞子、菊花分別浸洗；青邊鮑去腸頭，洗擦乾淨後氽水；紅蘿蔔去皮，切塊。
2. 青邊鮑魚、紅蘿蔔和薑片先用 7 碗水煮 2 小時，加入杞子、菊花，再煮 5 分鐘，調味即成。

食療功效

養陰固腎、益精明目。

飲食宜忌

本品滋補美味，老少可服。對手術恢復期出現目赤腫脹、視物昏花、血壓不穩、夜尿頻、氣虛哮喘、高血糖症狀及癌症患者有益。但痛風尿酸高者忌服。

認識主料

鮑魚

鮑魚肉中含有一種被稱為「鮑素」的成分,能夠破壞癌細胞必需的代謝物質,故有抗癌作用。青邊鮑宜整隻一齊煲煮,煮好才切片食,肉就不會韌。

鮑魚的外殼即中藥「石決明」,有平肝清熱、明目去翳的功效,所以如用鮮活的九孔鮑代替青邊鮑(大約 7-8 隻),最好連殼一齊煲。

菊花

菊花能散風清熱,平肝明目。菊花有極佳發散解熱之效,因此常用於外感風熱、畏寒、微汗等感冒初期症狀,對預防感冒也有效。但菊花性寒,體質偏寒者慎用。

淮杞花膠燉土雞

Double-steamed chicken soup with
Huai Shan and fish maw

材料（2人量）

淮山	38 克
杞子	6 克
浸發花膠	100 克
紅棗	6 粒
陳皮	1 塊
土雞	半隻

調味料

海鹽	1/4 茶匙

做法

1. 淮山、杞子浸洗，陳皮浸軟、去瓤；花膠汆水；紅棗去核；土雞斬大塊汆水。

2. 全部材料放入燉盅內，注入 3 碗開水，燉 3 小時後調味即成。

食療功效

健脾補血、滋補強壯。

飲食宜忌

本品鮮甜味美，老少可食。對手術恢復期間出現神疲乏力、
貧血、面色蒼白、腰膝痠軟、瘀血未散者有益。但脾胃消化
力弱者宜少食花膠，以免難以消化。

認識主料

土雞

是指放在山野林間、果園的肉
雞，街市偶有售賣，土雞的營養
較一般雞隻高，適合體弱消瘦、
免疫力低、記憶力下降、貧血、
水腫及發育遲緩的兒童食用；亦
可用有機雞隻代替。但外感發
熱、血脂高者不宜食。

花膠

含豐富的蛋白質、磷質及鈣
質，對肺腎虛弱，貧血等均有功
效。選購以色澤微黃油潤、無破
損、無白斑者為佳。

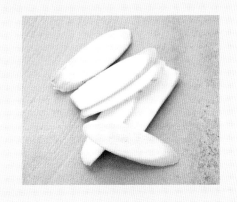

乾淮山

一般煮餸入饌以新鮮淮山為佳，它的補胃
滋陰生津功效較強；藥用則宜採用乾淮
山，它的健脾止瀉功效較佳。
一般人士都可食用，但便秘者宜少食。

茶樹菇馬蹄甘筍瘦肉湯

Lean pork soup with tea tree mushrooms,
water chestnuts and carrot

▎材 料 （ 2 人 量 ）

茶樹菇	20 克
馬蹄	6 粒
杏仁	10 克
甘筍	1 條
瘦肉	150 克

▎調 味 料

海鹽	1/4 茶匙

▎做 法

1. 茶樹菇浸洗，去蒂；馬蹄、甘筍去皮，切塊；杏仁沖洗；瘦肉切片，汆水。
2. 全部材料用 1 公升水煮個半小時，調味即成。

認識主料

茶樹菇

它有健脾止瀉、補腎滋陰的功效，對尿頻、腎虛、水腫、風濕等病症有獨特的食用療效，能緩解小兒尿床，並能提高人體免疫力，增強抗病能力。它含有人體所需 17 種氨基酸和十多種礦物質微量元素及抗癌多糖，其營養價值和保健功效均高於其他食用菌。但茶樹菇屬動風食物，過敏體質者忌食。

宜挑選味道清香無霉味、菇身比較粗大、淡棕色者。

食療功效

健脾補腎、清熱抗癌。

飲食宜忌

本品清甜好味，老少可飲。對手術恢復期出現腎虛、小便不利、尿頻、水腫、肺弱、氣喘痰多，以及對癌症康復病人有益。但對菇類敏感者不宜。

白果百合鷓鴣湯

—— 手術恢復期滋補強身湯水 ——

材料（2人量）	
白果	15 粒
百合	30 克
黃豆	30 克
生薑	3 片
紅棗	4 粒
鷓鴣	1 隻

調味料	
海鹽	1/4 茶匙

做法

1. 白果去芯，沖洗；百合、黃豆分別浸洗；紅棗去核；鷓鴣劏後洗淨，汆水。
2. 全部材料用 7 碗水煮 2 小時，調味即成。

滋補五臟、養肺安神。

本品滋潤美味，老少皆宜。對手術恢復期出現的肺氣弱、容易咳喘、夜尿多，精神不振、失眠多夢者有益。任何手術患者可服，但痛風患者不宜用黃豆。

白果

有收斂、化痰、止咳、定喘、止尿、止帶等功效。因果肉內含有氫氰酸，故不能生食，而且必須去芯才可減少毒性，成人每次食 15 粒為限。

鷓鴣

鷓鴣肉含有豐富的蛋白質、脂肪，且含有人體必需的18種氨基酸和較高的鋅、鍶等微量元素，具有壯陽補腎、強身健體的功效。民間把鷓鴣作為健脾消疳積的良藥，治療小兒厭食、消瘦、發育不良效果顯著。一般人士可食。

百合

百合能養陰潤肺，清心安神。對陰虛久咳、痰中帶血、虛煩驚悸、失眠多夢、精神恍惚有幫助。百合對白細胞減少症有預防作用，能升高血細胞，對化療及放射性治療後細胞減少症有治療作用。但風寒咳嗽、脾虛便溏者忌食。

青欖金羅漢果水

Luo Han Guo tea with green olives

材料（2人量）

青欖	8 粒
金羅漢果	半個

做法

1. 青欖洗淨，用刀背略拍鬆；羅漢果切碎。
2. 將材料用 5 碗水大火煮半小時即可供服。

食療功效

清肺利咽、生津止渴。

飲食宜忌

本品清甜，老少皆宜。對任何癌症電療、化療後咽乾喉燥、口苦、吞咽困難、大便秘結者皆有幫助。但脾胃虛寒者需加陳皮及生薑才可服用。

羅漢果

羅漢果有良好的利咽抗癌作用。傳統啡黑色的羅漢果經過高溫烘焙，會有點煙燻味。「金羅漢果」用低溫（攝氏 40-50 度）經 4-5 小時烘乾，能保留原果營養和味道，蜂蜜般清甜。它有清肺潤腸、利咽喉、降血糖、生津止渴等功效。可治喉痛、聲沙、氣管炎、哮喘、咳嗽、胃熱、便秘、急性扁桃體炎等症。但脾肯虛寒者不宜。

金羅漢果在大型中藥店或售賣烘焙生果乾的專門店有售。

剝開後的金羅漢果

青欖

橄欖的果實從幼到成熟，總是呈青綠色，故俗名青果。在果品中，橄欖風味獨特，初入口苦澀，稍嚼後轉為清香，滿口生津。青欖功能生津液，除煩熱，開胃降氣，清咽止渴，解毒醒酒。能解一切魚鱉之毒。胃寒者宜少食。

竹蔗茅根薏米水

Job's tears tea with sugarcane and lalang grass rhizome

食療功效

---| 食療功效 |---

清熱解毒、利水消腫。

---| 飲食宜忌 |---

本品清甜，對癌症患者出現口渴咽乾、煩躁口苦；或其他癌症電療期間喉嚨潰爛、吞咽困難者很適合，對白血病（血癌）出血傾向屬熱症亦很有益。但如患者體質偏寒及夜尿多者不宜。孕婦亦不宜吃薏米。

---| 認識主料 |---

竹蔗茅根

竹蔗茅根水是很普遍應用的保健涼茶，但並不適合所有人士的體質，尤其是體質偏寒者。超市賣的瓶裝竹蔗茅根水含糖量過高，常服反而對健康不利。

▌材料（2人量）

竹蔗	120 克
鮮茅根	120 克
生薏米	30 克

▌做法

1. 竹蔗洗淨，破開；鮮茅根洗淨，切段；薏米浸洗。
2. 燒熱 6 碗水，將材料放入後用大火煮滾，改用中火煮 40 分鐘成 2-3 碗即成。

Grape and lotus root juice

葡萄藕汁飲

▌材 料（ 2 人 量 ）

鮮榨葡萄汁	150 克
鮮蓮藕	200 毫升

▌做 法

1. 將新鮮葡萄洗淨，連皮、連核一齊榨汁；鮮蓮藕去皮洗淨，榨汁。
2. 將兩種汁放入煲內，加 100 毫升水，慢火煮滾即成，待涼才飲。

認識主料

鮮葡萄

能補氣、養血、健脾、強心。所含微量元素白藜蘆醇可防止細胞癌變，阻止癌細胞擴散。而鮮榨葡萄汁可以幫助器官移植手術患者減少排異反應，促進早日康復。

食療功效

養血益氣、涼血袪瘀。

飲食宜忌

本品清潤，老少皆宜，一般人均可服。對癌症電療、化療後出現出血現象、胃潰瘍、胃口不振、貧血體虛者有益，尤其是婦女卵巢癌患者可常服。

Tips 小貼士

蓮藕以粗壯者為佳，用蓮藕第二節榨汁最好。揀選蓮藕，皮要黃褐色，肉肥厚而白，如皮色發黑，有異味者不宜食用。

菜乾紅蘿蔔鴨腎湯

Lean pork soup with dried Bok Choy, carrot and dried duck gizzards

材料（2～3人量）

白菜乾	60 克
紅蘿蔔	1 條
瘦肉	100 克
陳鴨腎	2 個
蜜棗	2 粒

調味料

海鹽	1/4 茶匙

做法

1. 白菜乾浸洗淨，切段；紅蘿蔔去皮，切塊；瘦肉洗淨切片，與陳鴨腎一同汆水。
2. 全部材料用 1 公升水煮個半小時，調味即成。

食療功效

滋陰潤燥、養胃消食。

飲食宜忌

本品鮮甜可口，老少皆宜。對胃癌屬胃陰不足，其他癌腫電療期間和化療期間口乾口渴、消瘦、不思飲食、食難消化等均有幫助。任何癌症患者可服。

認識主料

白菜乾

能清胃熱、養胃陰。如大便秘結，影響腹部不適，心情煩躁等情況，或內臟有熱，引起頭痛、咽乾喉痛、胸骨或肋骨作痛等，都可以用白菜乾煲湯作舒緩。宜選購有機菜檔，乾身有香氣的有機菜乾。

舞茸冬菇杞子瘦肉湯

Lean pork soup with
mushrooms and Goji berries

材料（2～3人量）

舞茸菇	1 朵
冬菇	4 朵
杞子	3 克
紅棗	6 粒
陳皮	1 塊
瘦肉	200 克

調味料

海鹽	1/4 茶匙

做法

1. 舞茸菇、冬菇同浸洗，冬菇去蒂；陳皮、杞子浸洗，陳皮去瓤；紅棗去核；瘦肉切片，出水。
2. 全部材料用 1 公升水煮 1 小時，調味即成。

食療功效

補虛固表、益腎抗癌。

飲食宜忌

本品清香味美，老少皆宜。對抑制癌細胞的生長，減輕癌症電療、化療後對身體的副作用，提高機體免疫能力頗具療效。但對菇類敏感者忌服。

認識主料

舞茸菇、冬菇

舞茸菇、冬菇都是保健營養品，能增強人體免疫力，同時對患糖尿病、高血壓、肝病的人士有益。浸泡舞茸、冬菇的水不要倒掉，用來煲湯十分鮮美。

肉絲花膠粥

Congee with shredded pork and fish maw

材料（2人量）

梅頭瘦肉	50 克
浸發花膠	30 克
白米	60 克
薑	2 片
蔥粒	1 小撮

醃肉料

鹽、胡椒粉、生粉	各少許

調味料

海鹽	1/4 茶匙

做法

1. 瘦肉切絲，用醃料醃半小時；浸發花膠切絲；白米洗淨用少許油、鹽略醃。
2. 白米、薑片用 6 碗水煮成濃稠適度的粥，加入瘦肉絲、花膠絲煮至滾起，加入調味，灑上蔥粒即成。

食療功效

補益氣血、滋腎健脾。

飲食宜忌

本品香滑補益，老少皆宜。對婦女宮頸癌、卵巢癌等生殖系統癌症，以及其他癌腫手術後、電療、化療期間和治療後，出現消瘦虛弱、不思飲食皆有裨益。但脾濕痰多者不宜。

認識主料

花膠

花膠功能滋陰養血、固腎培精，對肺腎虛弱、貧血等甚為有益。更可令人消除疲勞，對外科手術病人傷口復元有幫助。花膠首要有厚度，煲起來不腥、不潺和不溶化，食時脍滑又爽口，才算佳品。但脾胃虛寒及消化力弱者宜少食。

白飯魚瑤柱粥
Whitebait congee with dried scallops

材料（2人量）

白飯魚	200 克
瑤柱	2-3 粒
白米	60 克
薑絲、葱花	各 1 湯匙

調味料

海鹽	1/4 茶匙

做法

1. 白飯魚洗淨，瀝乾；瑤柱浸軟，拆絲；白米洗淨，用少許油、鹽略醃。
2. 白米、瑤柱、薑絲用6碗水煮成濃稠適度的粥，加入白飯魚及調味，滾起，灑入葱花即成。

食療功效

健脾滋腎、補肺益陰。

飲食宜忌

本品香滑，老少皆宜。對胃癌、肺癌陰液不足，以及各種癌症患者電療期間或治療後陰虛內熱，口乾煩渴、消瘦食少、骨蒸潮熱、乾咳無痰等有幫助。但痛風者不宜。

認識主料

白飯魚

功能健胃益肺、補虛損，含豐富的蛋白質及多種維生素，是癌症患者補虛損的良好食物。坊間傳聞有假的白飯魚，白飯魚在汆水後，魚肉不會散掉，入口鬆軟腍滑有少許爽口才是正貨。

Dace ball soup with tomato and tofu

番茄豆腐魚丸湯

▌材料（2人量）

番茄	120 克
豆腐	1 磚
鯪魚滑	100 克
芫茜	1 棵

▌調味料

海鹽	1/4 茶匙

▌做法

1. 番茄去皮，切塊；豆腐沖洗切塊；鯪魚滑做成一粒粒魚丸；芫茜去根，洗淨切段。
2. 燒熱4碗水，加入番茄、豆腐煮滾，再加入魚丸滾10分鐘，加調味，灑入芫茜即成。

食療功效

清潤生津、開胃消食。

飲食宜忌

本品清香醒胃，老少皆宜。對各種癌症電療、化療期間和治療後胃津不足，食欲不振、口乾渴飲等均有幫助。如患者伴有嘔吐，可加少量薑汁或胡椒粉同用。

一般人士可服，痛風症患者宜少食豆腐。

認識主料

番茄

番茄能健胃消食、生津止渴。它含豐富的營養成分和多量維生素C，對煩躁、虛火上升、壓力大、睡不安寧都有不錯的緩解作用。番茄越紅，所含茄紅素越多；故宜選購顏色鮮紅、皮薄、底部平滑不要尖起的番茄。

電療、化療期飲食調理

Nutritious pumpkin soup

(makes 2 to 3 servings) Ref. p.014

▌Ingredients

150 g pumpkin
1 onion
1 potato
100 g spinach
60 g dried red kidney beans

▌Seasoning

1/4 tsp sea salt

▌Method

1. Peel pumpkin and potato. Rinse well and cut into chunks. Peel onion and cut into pieces. Soak and rinse red kidney beans in water. Cut off the roots of spinach. Rinse and cut into short lengths.
2. Boil 1 litre of water in a pot. Put in pumpkin, onion, potato and red kidney beans. Boil for 1 hour. Add spinach and season with sea salt. Boil for 5 more minutes. Serve both the soup and the solid ingredients.

▌Indications and contraindications

This soup is nutritious and is good for all ages. Those suffering from restlessness, thirst, high blood pressure, subconjunctival haemorrhage, Spleen- and Stomach-Asthenia, or poor digestive functions would benefit from serving this soup before or after surgery. Generally speaking, it is suitable for everyone. However, onion and spinach contain anti-coagulants. Patients should refrain from eating too much onion or spinach 2 to 3 days before surgery, and the whole week after.

Kudzu red date soup with small red beans and hyacinth beans

(makes 2 to 3 servings) Ref. p.016

▌Ingredients

250 g kudzu
30 g small red beans
30 g hyacinth beans
4 dried shiitake mushrooms
1 piece dried tangerine peel
6 red dates

▌Seasoning

1/4 tsp sea salt

▌Method

1. Peel the kudzu. Rinse and cut into chunks. Soak shiitake mushrooms in water till soft. Cut off the stems. Soak and rinse small red beans, hyacinth beans and dried tangerine peel in water. De-seed the red dates.
2. Put all ingredients into a pot. Add 1.2 litres of water. Boil for 1 1/2 hours. Season with sea salt. Serve.

▌Indications and contraindications

This soup is rich and tasty. It is good for all ages. Those suffering from tense muscles in the neck and shoulders, restlessness, depression, shortness of breath, or low spirits may consume this soup before or after surgery. Generally speaking, it is suitable for everyone.

Burdock detox soup
(makes 2 to 3 servings) Ref. p.019

Ingredients

150 g fresh burdock
6 small dried shiitake mushrooms
60 g carrot
100 g konjac noodles
150 g spinach
2 bowls vegetarian stock

Seasoning

1/4 tsp sea salt

Method

1. Scrub and rinse the burdock well without removing the skin. Slice thinly. Soak mushrooms in water till soft. Cut off the stems. Peel carrot and cut into chunks. Cut off the roots of the spinach. Rinse and cut into short lengths.
2. Boil the stock in a pot. Add 500 ml of water. Put in burdock, shiitake mushrooms and carrot. Boil for 40 minutes. Put in konjac noodles and spinach. Boil for 10 more minutes. Season with sea salt. Serve both the soup and the solid ingredients.

Indications and contraindications

This soup slims the body, detoxifies and beautifies the skin. This is suitable for those with weight problem, high blood pressure, diabetes or stroke before or after surgery. However, those with Asthenia-Coldness in the Spleen or Stomach meridians, and those suffering from diarrhoea or loose stool should avoid. Those taking blood thinner should not consume spinach before or after surgery. You may replace spinach with white cabbage.

Soybean soup with gingkoes, dried tofu skin and water chestnuts
(makes 2 to 3 servings) Ref. p.022

Ingredients

15 gingkoes
1 dried tofu skin
6 water chestnuts
60 g dried soybeans
3 g dried tangerine peel

Seasoning

1/4 tsp sea salt

Method

1. Shell gingkoes. Peel and core them. Rinse the dried tofu skin. Peel and rinse water chestnuts. Soak and rinse soybeans and dried tangerine peel in water.
2. Put all ingredients into a pot. Add 1 litre of water. Bring to the boil. Turn to low heat and simmer for 45 minutes. Season with sea salt. Serve both the soup and the solid ingredients.

Indications and contraindications

This soup is mild in nature without being too Cold or Hot from Chinese medical point of view. It is nutritious and easy to digest and absorb. Those suffering from Lung- and Kidney-Asthenia, and cough with much phlegm before or after surgery may feel free to consume. It is suitable for everyone in general.

Lotus root soup with carrot and peanuts

(makes 2 to 3 servings) Ref. p.024

▊ Ingredients

250 g lotus root
1 carrot
50 g peanuts
4 dried shiitake mushrooms
3 slices ginger
6 red dates

▊ Seasoning

1/4 tsp sea salt

▊ Method

1. Peel lotus root and carrot. Rinse and cut into chunks. Soak and rinse peanuts and shiitake mushrooms in water. Cut off the stems of the mushrooms. De-seed the red dates.
2. Boil 1 litre of water in a pot. Put in all ingredients. Boil for 1 hour. Season with sea salt. Serve the soup together with the solid ingredients.

▊ Indications and contraindications

This soup is sweet and delicious. Those suffering from Blood- and Qi-Asthenia, pale complexion, poor appetite, mental and physical exhaustion would benefit from serving this soup before or after surgery. However, those with indigestion should not eat too much of the solid ingredients.

Seaweed tofu soup with enokitake mushrooms

(makes 2 to 3 servings) Ref. p.026

▊ Ingredients

3 g dried seaweed
2 cubes tofu
1 pack enokitake mushrooms
1 tbsp shredded ginger
1 tbsp finely chopped spring onion

▊ Seasoning

1/4 tsp sea salt

▊ Method

1. Rinse seaweed and tofu in water. Cut tofu into small pieces. Cut off the roots of the enokitake mushrooms. Rinse and set aside.
2. Boil 600 ml of water in a pot. Put in seaweed, tofu, enokitake mushrooms and ginger. Bring to a boil. Season with sea salt and sprinkle with spring onion. Bring to the boil again. Serve.

▊ Indications and contraindications

This soup is velvety and yummy. It is good for all ages. Those with swollen thyroid, lymph node tuberculosis on the neck, high blood pressure, high blood triglycerides, or coronary heart disease would benefit from serving this soup before or after surgery. It is also suitable for cancer patients during radiotherapy or chemotherapy. However, those with Spleen- or Stomach-Asthenia and Coldness, or those with Qi- and Blood Asthenia should consume in strict moderation.

Egg drop soup with asparagus and white fungus

(makes 2 to 3 servings) Ref. p.028

▌Ingredients

200 g asparaguses
6 g dried white fungus
3 g dried Goji berries
2 egg whites
3 bowls vegetarian stock
1 tbsp caltrop starch slurry

▌Seasoning

1/4 tsp sea salt

▌Method

1. Rinse the asparaguses. Peel off the tough skin. Dice them. Soak white fungus in water till soft. Cut off the yellowish root. Chop finely. Soak and rinse Goji berries in water. Whisk the egg whites.
2. Boil the stock in a pot. Put in white fungus and cook for 20 minutes. Put in asparaguses and Goji berries. Boil for 5 minutes. Pour in caltrop starch slurry slowly while stirring continuously. Stir in whisked egg whites at last. Season with sea salt. Cook until the egg whites float. Turn off the heat. Serve.

▌Indications and contraindications

This soup is delicious and is good for all ages. Those with high blood pressure, coronary heart disease, digestive problems, anaemia, arthritis, oedema, neuritis, or obesity and cancer patients would benefit from serving this soup before or after surgery. However, gout patients should not consume.

Coconut soup with yellow ear and almonds

(makes 2 to 3 servings) Ref. p.031

▌Ingredients

1 copra (dried coconut kernel)
10 g yellow ear fungus
20 g sweet almonds
30 g peanuts
3 g dried Goji berries
6 red dates

▌Seasoning

1/4 tsp sea salt

▌Method

1. Cut the copra into chunks. Rinse well. Soak yellow fungus in water till soft. Cut off the root. Then cut into small florets with scissors. Rinse almonds. Soak and rinse peanuts and Goji berries separately. De-seed the red dates.
2. Boil 1 litre of water. Put in all ingredients. Boil for 1 hour. Season with sea salt. Serve.

▌Indications and contraindications

This soup is nourishing without being too greasy or filling. It is good for all ages. Those suffering from shortness of breath due to weak lungs, stomach or duodenum ulcers, insufficient body fluid secretion, dull and yellow complexion, or dry and wrinkly skin may feel free to serve this soup before or after surgery. However, those with fever due to influenza should avoid.

Bamboo fungus soup with day lily and white fungus

(makes 2 to 3 servings) Ref. p.035

Ingredients

6 pieces bamboo fungus
6 g day lily flowers
5 g dried white fungus
8 g dried Goji berries
2 tbsp frozen sweet corn kernels
10 chestnuts (shelled)

Seasoning

1/4 tsp sea salt

Method

1. Rinse and soak bamboo fungus, day lily flowers and white fungus in water until soft. Cut off the roots. Tie each day lily flower into a knot. Soak and rinse Goji berries in water. Thaw frozen corn kernels and rinse well. Blanch chestnuts in boiling water for a while. Peel them.
2. Boil 800 ml of water in a pot. Put in all ingredients. Boil for 30 minutes. Season with sea salt. Serve both the soup and the solid ingredients.

Indications and contraindications

This soup is light and sweet in taste without greasiness. Those who suffer from Asthenia-Heat with Dryness-Fire, insomnia and restless accompanied by Asthenia, dry throat and bitterness in the mouth, poor eyesight, depression, weight problem, or high blood pressure would benefit from serving it before or after surgical operations. However, white fungus contains anti-coagulants, so that patients should refrain from consuming any white fungus 2 to 3 days before surgery and the whole week after. You may use shiitake mushrooms instead.

Button mushroom soup with chayote and cashew nuts

(makes 2 to 3 servings) Ref. p.036

Ingredients

2 chayotes
50 g cashew nuts
60 g button mushrooms
2 slices ginger

Seasoning

1/4 tsp sea salt

Method

1. Peel and cut chayotes into chunks. Rinse the cashew nuts. Put off the stems of the button mushrooms. Rinse and drain well.
2. Boil 800 ml of water in a pot. Put in all ingredients. Boil for 30 minutes. Season with sea salt. Serve both the soup and the solid ingredients.

Indications and contraindications

This soup is nutritious and tasty. It is good for all ages. Those suffering from poor Qi flow in the Liver meridian, abdominal pain due to Qi congestion in the Liver or Stomach, high blood pressure, high cholesterol level or compromised brain power before or after surgery may feel free to serve this soup. Generally speaking, it is suitable for everyone.

Conch soup with green papaya and peanuts

(makes 2 to 3 servings) Ref. p.039

Ingredients

1 small green papaya
50 g peanuts
100 g frozen conch
3 dried figs
150 g lean pork

Seasoning

1/4 tsp sea salt

Method

1. Peel and de-seed the papaya. Cut into chunks. Soak and rinse peanuts in water. Drain. Thaw the frozen conch and blanch in boiling water. Drain. Slice the pork. Blanch in boiling water. Drain.
2. Put all ingredients into a pot. Add 1 litre of water. Boil for 1 hour. Season with sea salt. Serve the soup with the solid ingredients.

Indications and contraindications

This soup is flavourful and tasty. It is good for all ages. Those suffering from gastritis, gastric ulcer, duodenum ulcers, or restlessness due to Asthenic-Heat would benefit from serving it before or after surgery. Generally speaking it is suitable for everyone. Only pregnant women should refrain from eating papaya as it may trigger uterine contractions. Alternatively, you may use dried conch in place of frozen conch.

Double-steamed sea cucumber soup with fresh yam and Shi Hu

(makes 2 to 3 servings) Ref. p.042

Ingredients

150 g fresh yam
50 g fresh Shi Hu (Dendrobium stems)
5 g dried white fungus
1 piece chicken breast
2 rehydrated sea cucumber
4 black dates
2 slices ginger

Seasoning

1/4 tsp sea salt

Method

1. Peel fresh yam. Rinse well. Cut into pieces. Rinse the Shi Hu and cut into short lengths. Soak white fungus in water till soft. Cut off the yellowish root. Cut chicken breast into pieces. Blanch in boiling water. Drain. Cut sea cucumber into pieces. Blanch in boiling water. Drain.
2. Put all ingredients into a double-steaming pot. Add 500 ml of boiling water. Double-steam for 3 hours. Season with sea salt. Serve.

Indications and contraindications

This soup is nourishing and tasty. It is good for all ages. Those suffering from insufficient body fluid secretion, Qi- and Blood-Asthenia, physical and mental exhaustion, poor eyesight, or constipation before or after surgery, or after radiotherapy or chemotherapy would benefit from serving this soup. However, those with fever due to influenza, accumulate of Phlegm-Dampness, or those suffering from diarrhoea should not consume.

Beef soup with onion and tomato
(makes 2 to 3 servings) Ref. p.044

▍Ingredients

1 onion
3 tomatoes
300 g beef
2 slices ginger

▍Seasoning

1/4 tsp sea salt

▍Method

1. Peel tomatoes. Rinse and cut into chunks. Peel and slice onion. Slice the beef and blanch in boiling water. Drain.
2. Put all ingredients into a pot. Add 800 ml of water. Boil for 30 minutes. Season with sea salt. Serve.

▍Indications and contraindications

This soup boosts body energy and immune system. It is good for all ages. Those suffering from high blood pressure, diabetes, coronary heart disease, obesity or cancer would benefit from serving this soup before or after surgery. However, beef may trigger skin problems among some people so that those with eczema or skin problems should consume beef in strict moderation. Onion also contains anti-coagulants. Thus, patients should not eat any onion 2 to 3 days before surgical operations and the whole week after.

Grass carp soup with Chinese marrow and spirulina
(makes 2 to 3 servings) Ref. p.046

▍Ingredients

1 Chinese marrow
5 g dried spirulina
1 grass carp fillet
1/2 tbsp shredded ginger

▍Seasoning

1/4 tsp sea salt

▍Method

1. Scrape off the skin of the Chinese marrow with a knife. Cut into chunks. Rinse the spirulina. Rinse the grass carp fillet and slice thinly.
2. Boil 800 ml of water in a pot. Put in Chinese marrow and ginger. Boil for 15 minutes. Put in spirulina and grass carp. Cook for 10 more minutes. Season with sea salt. Serve.

▍Indications and contraindications

This soup is tasty and flavourful. It is good for all ages. Those suffering from Qi- or Blood-Asthenia, malnutrition, high blood pressure, high blood triglycerides, diabetes, iron deficiency anaemia or tumour may feel free to serve this soup before or after surgery. Generally speaking, it is suitable for everyone.

Double-steamed teal soup with cordyceps spores and fish maw

(makes 2 to 3 servings) Ref. p.048

▌Ingredients

6 g cordyceps spores
150 g rehydrated fish maw
5 dried Goji berries
10 g dried longans
1 chilled teal
2 slices ginger

▌Seasoning

1/4 tsp sea salt

▌Method

1. Rinse and soak cordyceps spores, dried Goji berries and dried longans separately in water. Dress the teal and rinse well. Chop into pieces. Blanch teal and fish maw in boiling water together. Drain.
2. Put all ingredients into a double-steaming pot. Pour in 3 bowls of boiling water. Double-steam for 3 hours. Season with sea salt. Serve.

▌Indications and contraindications

This soup is fragrant and tasty. It is good for all ages. Those suffering from Yin-Asthenia with accumulated Heat, restlessness, diabetes, high blood pressure, shortness of breath due to weak lungs, or any kind of cancer may feel free to serve this soup before or after surgery. However, those with fever due to influenza or indigestion should not consume.

Pearl clam soup with walnuts and lotus seeds

(makes 2 to 3 servings) Ref. p.051

▌Ingredients

30 g walnuts (shelled)
20 g dried lotus seeds
20 g fox nuts
5 g Goji berries
3 dried pearl clams (shelled)
250 g lean pork

▌Seasoning

1/4 tsp sea salt

▌Method

1. Slice the pork. Blanch in boiling water. Drain. Blanch the pearl clams in boiling water. Drain. Rinse the rest of the ingredients.
2. Put all ingredients into a pot. Add 1 litre of water. Boil for 1 1/2 hours. Season with sea salt. Serve.

▌Indications and contraindications

This soup is rich and delicious. It is good for all ages. Those suffering from Liver- and Kidney-Asthenia, poor eyesight, low spirits or frequent urinations at night would benefit from serving this soup before or after surgery. However, those with fever due to influenza should not consume.

Pond loach soup with tofu

(makes 2 to 3 servings) Ref. p.054

▌Ingredients

2 cubes firm tofu
250 g pond loach
2 slices ginger
2 sprigs spring onion (cut into short lengths)

▌Seasoning

1/4 tsp sea salt

▌Method

1. Cut tofu into pieces. Put pond loach into a muslin bag and tie well. Blanch the pond loach in boiling water for 2 to 3 minutes. Take the fish out and scrub off the slime. Then remove innards and fry in a little oil until fragrant.
2. Put tofu, ginger and pond loach into a pot. Add 800 ml of water. Boil for 30 minutes. Sprinkle with spring onion. Season with sea salt. Serve.

▌Indications and contraindications

This soup is sweet and tasty. It is good for all ages. Those with hepatitis, liver cancer, or ascites due to liver cirrhosis would benefit from serving this soup before or after surgery. However, those suffering from Spleen- or Liver-Asthenia or Coldness should add 1 tsp of whole white peppercorns to the soup.

Quail soup with dried cuttlefish and red kidney beans

(makes 2 to 3 servings) Ref. p.056

▌Ingredients

30 g dried red kidney beans
60 g dried cuttlefish
4 red dates
2 quails
3 slices ginger

▌Seasoning

1/4 tsp sea salt

▌Method

1. Soak and rinse red kidney beans in water. Soak and rinse dried cuttlefish in water. Dress the quail and rinse well. Blanch dried cuttlefish and quail together in boiling water. Drain. De-seed the red dates.
2. Put all ingredients into a pot. Add 1.2 litres of water. Boil for 1 1/2 hours. Season with sea salt. Serve.

▌Indications and contraindications

This soup is tasty and flavourful. It is good for all ages. It helps alleviate premature grey hair, dizziness and palpitations due to Blood-Asthenia. It is especially useful for those suffering from low white blood cell count after radiotherapy. However, those with fever, influenza or skin allergy should not consume.

Pork liver soup with day lily flowers and spinach

(makes 2 to 3 servings) Ref. p.058

Ingredients

6 g day lily flowers
200 g spinach
4 g dried Goji berries
250 g pork liver

Seasoning

1/4 tsp sea salt

Method

1. Soak day lily flowers in water till soft. Tie each into a knot. Cut off the roots of the spinach. Rinse and cut into short lengths. Soak and rinse Goji berries in water. Set aside. Rinse the pork liver. Slice and blanch in boiling water.
2. Put all ingredients into a pot. Add 600 ml of water. Bring to the boil and cook for 15 minutes. Season with sea salt. Serve the soup with the solid ingredients.

Indications and contraindications

This soup is fragrant and is good for all ages. Those with hepatitis, nervous prostration, swollen and aching breasts, night blindness, difficulty in passing urine, or bloody stool may feel free to serve this soup before or after surgery. However, spinach contains anti-coagulants so that those taking blood thinner, gout patients and those suffering from diarrhoea should avoid.

Pork rib soup with lotus seed, Fu Shen and dried longans

(makes 2 to 3 servings) Ref. p.060

Ingredients

30 g lotus seeds
20 g Fu Shen
15 g dried longans
30 g fox nuts
250 g pork ribs

Seasoning

1/4 tsp sea salt

Method

1. Rinse and soak lotus seeds, Fu Shen and fox nuts in water. Rinse dried longans in water. Drain. Blanch pork ribs in boiling water. Drain.
2. Put all ingredients into a pot. Add 1.2 litres of water. Boil for 1 1/2 hours. Season with sea salt. Serve.

Indications and contraindications

This soup is fragrant and tasty. It is good for all ages. Those suffering from anaemia, palpitations, insomnia, poor memory, excessive sweating due to Asthenia, or nervous prostration before or after surgery would benefit from serving this soup. However, those with constipation or influenza should not consume.

Soybean milk with mung bean and Job's tear puree
(makes 1 serving) Ref. p.062

▌Ingredients

30 g mung beans
30 g Job's tears
250 ml unsweetened black soybean milk

▌Method

1. Soak mung beans and Job's tears in water separately for half a day. Drain and transfer into a blender. Add black soybean milk. Blend until fine.
2. Pour the resulting mixture into a pot. Cook for 10 minutes. Serve.

▌Indications and contraindications

This drink energizes and strengthens the body, while reducing blood glucose and cholesterol levels. Those suffering from restlessness, thirst, difficulty passing stool or urine, or itchy skin would benefit from serving this drink before or after surgery. However, pregnant women and those with Spleen- or Stomach-Asthenia and Coldness should not consume.

Apple and veggie juice
(makes 1 serving) Ref. p.065

▌Ingredients

1 apple
1 carrot
1 potato

▌Method

1. Rinse apple, carrot and potato. Wipe dry and peel them. Cut them into pieces.
2. Put all ingredients into a slow juicer. Serve the juice blend on an empty stomach.

▌Indications and contraindications

Drink this juice right after it is squeezed. Do not make any extra for storage. Serve it once daily and serve continuously for 1 to 3 months. It helps fight cancer, protects the liver and boosts physical strength before and after surgical operation. However, those with kidney problem should consume in strict moderation. Otherwise, the high potassium content in the juice may put extra burden on your kidneys.

Water chestnut, lotus root and pear juice

(Makes 1 serving) Ref. p.068

Ingredients

6 water chestnuts
150 g lotus root
1 Ya-li pear

Method

1. Peel and rinse water chestnuts and lotus root. Finely chop them. Peel and rinse the pear. Cut in quarters. Remove the core and finely chop it.
2. Press all ingredients into a slow juicer. Pour the resulting juice blend into a pot. Boil for 5 minutes. Serve.

Indications and contraindications

This juice is sweet in taste. It helps alleviate insufficient body fluid secretion, dry mouth and throat, restlessness while feeling hot, dry and hard stool before or after surgical operations. It is especially beneficial to those receiving operations in the lungs. However, those with Coldness or Asthenia in the Spleen and Stomach, and those having watery stool should not consume.

Red date and kiwi tea

(makes 1 serving) Ref. p.070

Ingredients

2 kiwis
30 g red dates
3 g black tea leaves

Method

1. Rinse and peel the kiwis. Finely chop them. Rinse the red dates and de-seed them.
2. Put kiwis and red dates into a pot. Add 3 bowls of water and boil for 15 minutes. Put in the tea leaves. Turn off the heat and cover the lid. Let stand for 5 minutes. Strain and serve.

Indications and contraindications

You may drink this tea with an empty stomach to beautify your skin. It also helps promote peristalsis along the digestive tract and eliminates residual stool in the intestines. It is suitable for patients with cancer, cardiac vascular diseases, high blood pressure and poor appetite before or after the surgery. However, those suffering from renal failure, those allergic to kiwis, and infants especially should not consume.

Walnut and cashew milk with apple puree

(makes 1 serving) Ref. p.072

Ingredients

10 g walnuts (shelled)
10 g cashew nuts
1 apple
200 ml partly skimmed milk

Method

1. Peel and core the apple. Cut into strips.
2. Put walnuts, cashew nuts, apple and a little partly skimmed milk into a blender. Blend until fine.
3. Add the remaining milk. Stir well. Transfer into a pot and heat briefly. Serve.

Indications and contraindications

This drink is highly nutritious. It strengthens the body in general and defies aging. It is suitable for those with diabetes, cardiac vascular diseases or dementia before and after surgical operations. However, those suffering from diarrhoea or those with lactose intolerance should not consume.

Banana soymilk

(makes 1 serving) Ref. p.074

Ingredients

1 banana
200 ml soymilk (at room temperature)

Method

1. Peel the banana and cut into pieces. Put into a blender and add a little soymilk. Blend until smooth.
2. Pour the mixture into a serving glass. Add remaining soymilk. Mix well and serve.

Indications and contraindications

This drink is sweet and fragrant. Just make sure the soymilk is left to warm up to room temperature before serving. Do not serve this drink chilled. Those suffering from fever, restlessness, thirst, dry stool, constipation, cardiac vascular diseases, dry skin, or cough due to Heat accumulated in the Lungs may serve this drink before or after surgical operations. However, those with Spleen- or Stomach-Asthenia or Coldness, frequent urinations at night, acute or chronic nephritis should not consume.

Honey lemon passionfruit tea
(makes 1 serving) Ref. p.076

▌Ingredients

2 slices lemon
honey
1 passionfruit

▌Method

1. Soak the whole lemon in boiling water for a while. Rinse well to remove the wax and residual insecticide on the skin.
2. Wipe dry the lemon and slice it. Transfer into a glass jar. Pour honey over it to cover. Cover the lid and leave it in the fridge overnight.
3. Put 2 slices of honey lemon with a little honey in the jar into a cup. Slice a passionfruit and spoon the pulp and seeds into the cup. Add warm water. Stir well and serve.

▌Indications and contraindications

This tea boosts the immune system, regulates blood flow and alleviates depression. It is recommended to those with anxiety, nervousness, or poor appetite before and after surgical operations. However, those allergic to passionfruit should avoid. Those with poor digestion and excessive stomach acid should use with care.

Red date brown rice tea
(makes 1 serving) Ref. p.078

▌Ingredients

4 red dates
2 tbsp brown rice

▌Method

1. De-seed the red dates and slice them thinly. Fry brown rice in a dry wok over low heat for 7 minutes.
2. Put all ingredients into a teapot. Pour in boiling water. Cover the lid and let stand for 15 minutes. Serve.

▌Indications and contraindications

This tea is fragrant and is good for all ages. Those with high blood triglycerides, high blood cholesterol, anaemia, obesity, cardiac vascular diseases, deterioration of brain power or cancer may consume before or after surgical operations. However, those who suffer from impaired digestion after operation should only drink the soup without eating the brown rice.

Poached fish fillet in vinegar sauce

(makes 3 to 4 servings) Ref. p.083

▮ Ingredients

150 g frozen fish fillet
1/4 red bell pepper
1/4 yellow bell pepper
3 g cloud ear fungus
2 slices ginger
1 sprig coriander

▮ Marinade

1/6 tsp salt
ground white pepper
caltrop starch
rice wine

▮ Seasoning

2 tbsp rice vinegar
1 tbsp light brown sugar
1/4 tsp salt
2 tsp light soy sauce
ground white pepper
30 ml water

▮ Method

1. Thaw the fish fillet and slice it. Add marinade and mix well. Leave it for 30 minutes. Rinse the bell peppers and cut into pieces. Soak cloud ear fungus in water till soft. Cut off the root. Rinse coriander and finely chop it.
2. Boil a pot of water. Put in the sliced fish and turn off the heat. Leave it briefly. Drain and set aside the fish.
3. Heat some oil in a wok. Stir-fry ginger until fragrant. Add bell peppers and cloud ear fungus. Stir-fry till fragrant. Sizzle with wine and add seasoning. Bring to the boil. Put in the fish fillet from step 2 and coriander. Stir well and serve.

▮ Indications and contraindications

This dish is sour and sweet in taste. It is appetizing and good for all ages. Cancer patients and those with high blood pressure, high cholesterol level or angina pectoris who suffer from poor appetite, mental and physical exhaustion or constipation before or after surgery may serve this dish. However, those with bone fracture, sports injuries, and those taking blood thinner should not consume cloud ear fungus.

Four-red tea

(makes 1 serving) Ref. p.080

▮ Ingredients

30 g redskin peanuts
30 g red beans
6 g dried Goji berries
4 red dates

▮ Seasoning

1 tbsp light brown sugar

▮ Method

1. Rinse red beans, peanuts and Goji berries separately. De-seed the red dates.
2. Put all ingredients into a pot and add 5 bowls of water. Boil for 1 hour. Season with light brown sugar. Serve.

▮ Indications and contraindications

This tea is sweet and delicious. It is good for all ages. Patients suffering from Blood-Asthenia, pale complexion, oedema, insomnia or persistent hepatitis would benefit from serving this tea before and after surgical operations. However, diabetics and those with sports injuries or bruises should not consume.

Stir-fried shredded chicken with Chinese celery and water bamboo shoot

(makes 3 to 4 servings) Ref. p.086

Ingredients

2 chicken tenderloins
200 g water bamboo shoots
30 g Chinese celery
1 small segment carrot
1 tsp grated ginger

Marinade for chicken

salt
ground white pepper
rice wine
cornstarch

Seasoning

1/2 tsp salt
1 tsp oyster sauce
1/2 tbsp rice wine

Method

1. Rinse the chicken and cut into thick strips. Add marinade and mix well. Leave it for 30 minutes. Slice carrot and cut into floral shapes. Remove the tough outer leaves on water bamboo shoots. Rinse and cut into thick strips. Rinse Chinese celery and cut into short lengths.
2. Heat some oil in a wok. Stir-fry ginger until fragrant. Put in the chicken and toss until it turns white. Set aside.
3. Save some oil in the same wok. Stir-fry water bamboo shoots, Chinese celery and carrot for a while. Put the chicken back in. Sizzle with wine. Add seasoning and cook until the sauce reduces. Serve.

Indications and contraindications

This dish is aromatic with a crunchy texture. It is good for all ages. Those suffering from high blood pressure, diabetes, Dampness-Heat, jaundice, red swollen eyes that hurt may feel free to eat it before or after surgery. However, those with Asthenia-Coldness in the Spleen or Stomach and those with renal, bladder or ureteral stones should avoid.

Steamed pumpkin with tofu pork balls

(makes 3 to 4 servings) Ref. p.088

Ingredients

250 g pumpkin
1/2 tbsp Dong Cai (salted cabbage with garlic)
1/2 cube firm tofu
1 tsp grated ginger
2 tsp finely chopped spring onion
250 g ground pork

Marinade

1 egg white
1/2 tsp salt
1/2 tbsp light soy sauce
1 tsp cornstarch

Seasoning

1 tsp oyster sauce
1/2 tbsp light soy sauce

Method

1. Rinse the pumpkin. Peel and cut into pieces. Lay evenly on the bottom of a steaming dish.
2. Rinse Dong Cai in water and squeeze dry. In a mixing bowl, mix together Dong Cai, mashed tofu, grated ginger, spring onion, marinade and ground pork. Stir until sticky. Squeeze into meat balls. Arrange evenly over the bed of pumpkin on the steaming dish.
3. Steam over high heat for 20 minutes. Sprinkle with mixed seasoning. Serve.

Indications and contraindications

This dish is sweet and tasty. It is good for all ages. Those suffering from diabetes, cardiac vascular diseases or constipation due to Dryness in the intestines would benefit from consuming this dish before or after surgery. However, those with Qi stagnation and accumulated Dampness should consume in moderation.

Stir-fried ground pork with soybean sprouts

(makes 3 to 4 servings) Ref. p.090

Ingredients

250 g soybean sprouts
100 g ground pork
1 tbsp shredded red bell pepper
1 tbsp finely shredded spring onion

Marinade

light soy sauce
ground white pepper
cornstarch

Seasoning

1/4 tsp salt
1 tsp oyster sauce
sesame oil

Method

1. Add marinade to ground pork. Mix well. Rinse bean sprouts. Cut off the roots and finely chop them.
2. Fry soybean sprouts in a dry wok over high heat until dry and cooked through. Set aside.
3. Heat some oil in a wok. Fry the ground pork until lightly browned. Press with a spatula to break the pork into small bits. Add soybean sprouts, red bell pepper and seasoning. Toss until the pork is cooked through. Sprinkle with finely shredded spring onion. Toss briefly. Serve.

Indications and contraindications

This is a home-style dish that nourishes the Yin, and promotes the secretion of body fluid. It is good for all ages. Those with skinny build or poor appetite also benefits from serving this dish before or after surgery. This recipe is also good for those cancer patients who suffer from dry mouth and throat after radiotherapy or chemotherapy. However, gout patients should consume beans in strict moderation.

Stir-fried kale stems with chayote

(Makes 3 to 4 servings) Ref. p.092

Ingredients

1 chayote
100 g kale stems
3 g dried day lily flowers
1 small segment carrot
1 tsp grated ginger
30 ml stock
1 tbsp rice wine
1 tbsp caltrop starch slurry

Seasoning

1/2 tsp salt
1/4 tsp sugar

Method

1. Peel, de-seed and slice the chayote. Soak the day lily flowers in water till soft. Tie each into a knot. Rinse the kale stems and slice them. Slice carrot and cut into floret pattern.
2. Heat a wok and add some oil. Stir-fry ginger until fragrant. Put in kale, chayote and day lily flowers. Toss till fragrant. Sizzle with wine. Add seasoning and stock. Toss well. Stir in caltrop starch slurry. Toss evenly. Serve.

Indications and contraindications

This dish is light and refreshing. It is good for all ages. Those suffering from Qi- and Blood-Asthenia, or poor Qi flow in the Liver meridian before or after gallstone or intrahepatic biliary stone operation would benefit from eating it. However, those suffering from Stomach-Coldness with excessive secretion of saliva should consume in moderation.

Steamed Chinese marrow rings stuffed with ground pork

(makes 3 to 4 servings) Ref. p.103

Ingredients

1 large Chinese marrow
200 g ground pork
1 tbsp dried shrimps
1 tbsp finely chopped spring onion

Marinade for pork

1 egg white
salt
ground white pepper
cornstarch

Thickening glaze

2 tsp light soy sauce
1/2 tbsp rice wine
1/4 tsp sugar
1 tbsp caltrop starch slurry

Method

1. Scrape off the skin of the Chinese marrow with a knife. Cut into slices about 2-cm thick. Scoop out the seeds to make a ring. Blanch Chinese marrow in boiling water with a pinch of salt and a dash of oil for 30 seconds. Drain well.
2. Soak dried shrimps in water till soft. Drain. Finely chop them. Add dried shrimps and spring onion to the ground pork. Add marinade and stir until sticky.
3. Dust the insides of the Chinese marrow rings with some cornstarch. Stuff the rings with ground pork filling. Arrange on a steaming plate. Steam for 25 minutes. Drain any liquid on the plate. Mix the thickening glaze well and bring to the boil in a pan. Pour the glaze over the steamed Chinese marrow rings. Serve.

Indications and contraindications

This dish tastes refreshing and light. Those with mental exhaustion, poor appetite or dry throat, and those with Asthenia whose body cannot absorb the potent medicinal in health tonic, would benefit from serving this dish before or after surgery. Generally speaking, everyone may consume this dish.

Shredded omelette with assorted veggies

(makes 3 to 4 servings) Ref. p.094

Ingredients

3 eggs
1/2 bulb fennel
1 dried shiitake mushroom
1 small segment carrot
1 small segment cucumber

Seasoning for omelette

1/4 tsp salt
1 tsp sesame oil

Seasoning

1/2 tsp salt

Method

1. Whisk the eggs. Add salt and sesame oil. Whisk again to mix well. Rinse the bulb fennel and finely shred it. Soak the shiitake mushroom in water till soft. Cut off the stem and finely shred it. Peel and rinse the carrot. Finely shred it. Peel and rinse the cucumber. Finely shred it.
2. Heat a pan and add oil. Fry the egg into an omelette until browned on both sides. Let cool and shred it.
3. Save some oil in a pan. Put in fennel, shiitake mushrooms, carrot and cucumber. Stir until the liquid reduces. Put in the shredded omelette from step 2 and seasoning. Toss well. Serve.

Indications and contraindications

This dish is tasty and aromatic. It is good for all ages. Those suffering from poor appetite, Qi- and Blood-Asthenia, or memory loss may consume this dish before or after surgery. Generally speaking, it is suitable for everyone.

Stir-fried fresh yam with five-colour veggies

(makes 3 to 4 servings) Ref. p.096

Ingredients

250 g fresh yam
30 g red onion
30 g fresh pineapple
10 g green bell pepper
10 g red bell pepper
1 tsp grated ginger

Marinade for yam

1/4 tsp salt
1 tbsp caltrop starch

Seasoning

2 tsp rice vinegar
2 tsp light brown sugar
1 tbsp tomato paste
1/4 tsp salt
1/2 tbsp light soy sauce
cornstarch

Method

1. Peel and rinse the yam. Cut into thick slices. Add marinade and mix well. Leave it briefly. Peel and slice red onion. Dice pineapple finely. Cut bell peppers into pieces.
2. Heat a pan and add some oil. Fry yam until lightly browned. Set aside. Put all remaining ingredients into the pan. Toss until fragrant. Put the yam back in. Pour in the seasoning and cook until the sauce thickens. Serve.

Indications and contraindications

This dish is sour and appetizing. Those suffering from poor appetite, poor digestion, physical weakness, Spleen-Asthenia and diarrhoea before or after surgery would benefit from serving it. However, those with influenza or indigestion should not consume.

Braised sea cucumbers with shiitake mushrooms

(makes 3 to 4 servings) Ref. p.100

Ingredients

6 to 8 small dried shiitake mushrooms
2 rehydrated sea cucumbers
1/2 tbsp shredded ginger
6 sprigs Bok Choy (stems only)
1 bowl stock
1 tbsp rice wine
1 tbsp caltrop starch slurry

Seasoning

1/2 tsp salt
1/2 tsp sugar
2 tsp oyster sauce

Method

1. Soak the shiitake mushrooms in water till soft. Cut off the stems. Cut sea cucumbers into pieces. Blanch in boiling water. Drain.
2. Heat a little oil in a wok. Stir-fry ginger until fragrant. Add shiitake mushrooms and sea cucumbers. Fry until fragrant. Sizzle with wine. Add seasoning and stock. Bring to the boil and cook for 30 minutes.
3. Rinse the Bok Choy stems. While the mushrooms and sea cucumbers are cooking, blanch Bok Choy stems in boiling water with a dash of oil and a pinch of salt. Cook until Bok Choy are just done and still bright green. Drain. Arrange Bok Choy on the rim of a serving dish.
4. Add caltrop starch slurry to the mushrooms and sea cucumbers from step 2 while stirring continuously. Pour the resulting mixture on the serving dish. Serve.

Indications and contraindications

This dish is tasty and the mushrooms and sea cucumbers are chewy and gelatinous in texture. It is good for all ages and especially suitable for those with high blood pressure, high cholesterol level, tuberculosis, hepatitis, neuritis, haemophilia or cancer before or after surgery. However, those with obesity and blood stasis due to Phlegm-Dampness, diarrhoea or gout should avoid.

Stir-fried dried tofu with assorted vegetables

(makes 3 to 4 servings) Ref. p.098

▓ Ingredients

3 g wood ear fungus
1/2 carrot
30 g snow peas
3 cubes five-spice dried tofu
1 tsp grated ginger
1 tsp grated garlic
2 tsp Shaoxing wine
1 tbsp caltrop starch slurry

▓ Seasoning

1/2 tsp salt
1/2 tsp sugar
2 tsp oyster sauce

▓ Method

1. Soak wood ear fungus in water until soft. Cut off the stem and finely shred it. Peel and shred carrot. Tear the tough veins off the snow peas. Rinse well and finely shred them. Rinse the dried tofu and finely shred it.
2. Heat oil in a wok. Fry the dried tofu over low heat until golden. Set aside. Save some oil in the same wok. Heat it up and stir-fry garlic and ginger until fragrant. Add wood ear fungus, carrot and snow peas. Toss until cooked through.
3. Put the shredded dried tofu back in. Sizzle with wine and add the remaining seasoning. Toss again. Stir in caltrop starch slurry at last. Toss evenly and serve.

▓ Indications and contraindications

This dish is rich in dietary fibre that promotes peristalsis along the digestive tract and helps weight loss. This is a great dish for those with weight problem, high blood triglycerides, or high blood cholesterol before or after surgical operations. However, those taking blood thinning pills should not consume wood ear fungus because it contain anti-coagulant. Thus, they should refrain from consuming any wood ear fungus 2 to 3 days before and 1 week after surgery. For this recipe, they may use shiitake mushrooms in place of wood ear fungus.

Radish and dried tangerine peel tea

(makes 1 serving) Ref. p.111

▓ Ingredients

1 white radish
2 whole dried tangerine peel

▓ Method

1. Rinse the white radish. Peel and cut into chunks. Soak dried tangerine peel in water till soft. Rinse well.
2. Put ingredients into a pot. Add 5 bowls of water and bring to the boil. Cook for 30 minutes until the liquid reduces to 2 to 3 bowls. Serve.

▓ Indications and contraindications

Drink this tea in big gulps to stimulate peristalsis along the digestive tract and promote bowel movements. Drink it in tiny sips to activate bladder movement and the excretion of the residual anaesthetic and toxins in the body after surgery. It also helps dissipate the phlegm. It is suitable for all patients after surgery.

Wheat groats and black bean tea
(makes 1 serving) Ref. p.108

▌Ingredients

30 g wheat groats
30 g black beans with green kernels
10 g Fu Shen
8 dried longans (shelled and de-seeded)

▌Method

1. Rinse and soak wheat groats, black beans and Fu Shen in water separately. Rinse the dried longans.
2. Put all ingredients in a pot. Add 5 bowls of water. Boil for 1 hour. Serve.

▌Indications and contraindications

This tea is fragrant in taste. Those suffering from restlessness, anxiety, light sleep with many dreams, excessive sweating, pale complexion due to Blood-Asthenia and arthritic pain after surgery will benefit from serving this tea. However, those suffering from gout should refrain from consuming beans.

Red date and day lily tea
(makes 1 serving) Ref. p.116

▌Ingredients

5 g day lily flowers
4 red dates

▌Seasoning

1 tsp light brown sugar

▌Method

1. Rinse and soak the day lily in water. Rinse red dates and de-seed them. Then slice them.
2. Put day lily and red dates into a pot. Add 2 1/2 bowls of water. Bring to the boil and cook for 10 minutes. Add light brown sugar. Cook until sugar dissolves. Serve.

▌Indications and contraindications

This tea is delicious and sweet. It is suitable for patients suffering from general weakness, exhaustion, pale complexion, low blood pressure, difficulty in urinating or poor memory after surgery. Patients may feel free to consumer after any surgery.

Bitter buckwheat tea with Goji berries
(makes 1 serving) Ref. p.106

▌Ingredients

2 tbsp bitter buckwheat
1 tsp dried Goji berries

▌Method

1. Put bitter buckwheat and Goji berries into a teabag. Seal well. Put the teabag into a teapot. Rinse with boiling water once. Drain.
2. Pour in boiling water again. Cover the lid and leave it for 10 minutes. Serve.

▌Indications and contraindications

This tea is tasty and fragrant. Those suffering from high blood cholesterol, high blood pressure and high blood triglycerides would benefit from it. It also helps alleviate restlessness, insomnia, difficulty in urinating and pale complexion among patients after surgery. However, those with Cold bodily predisposition and those with frequent urination at night should consume in moderation.

Brown sugar lemonade

(makes 1 serving) Ref. p.114

▌Ingredients

1/2 fresh lemon

▌Seasoning

2 tsp light brown sugar

▌Method

1. Soak the whole lemon in boiling water briefly. Rinse again. This helps remove the wax and insecticides.
2. Cut the lemon in half. Slice half of the lemon and put it into a tea pot. Add light brown sugar and boiling water. Cover the lid and let stand for 5 minutes before serving.

▌Indications and contraindications

This drink is generally good for everyone. Those suffering from dry mouth, anxiety, low spirits, physical and mental exhaustion after surgery would especially benefit. Serve after any surgery. However, those with stomach ulcer or excessive stomach acid should not consume.

Egg drop soup with white fungus

(makes 2 servings) Ref. p.119

▌Ingredients

6 g white fungus
20 g fresh lily bulbs
1 egg
1 tsp dried Goji berries
3 bowls stock

▌Seasoning

1/2 tsp sea salt

▌Method

1. Soak white fungus in water till soft. Rinse and cut off the yellowish root. Finely chop it and set aside. Break the lily bulbs into scales. Rinse well. Whisk the egg.
2. Boil the stock and put in white fungus, lily bulbs and Goji berries. Cook for 30 minutes. Season with sea salt and stir in the egg. Bring to the boil again. Serve.

▌Indications and contraindications

This soup is delicious. It alleviates Liver-Asthenia without triggering Fire in the Liver meridian. Those who suffer from poor sleep quality, low spirits, Yin-Asthenia with overwhelming Fire, cough due to Heat in the Lungs, or blood in phlegm after surgery would benefit from serving it. Generally speaking, all patients may consume after operations. Only those taking blood thinner should avoid.

Tofu thick soup with laver and sweet corn

(Makes 2 servings) Ref. p.122

▌Ingredients

1 small pinch dried laver
1 tbsp frozen sweet corn kernels
2 tsp frozen green peas
1 box soft tofu
2 bowls stock
2 tbsp caltrop starch slurry
1 egg white

▌Seasoning

1/2 tsp salt

▌Method

1. Rinse the laver. Thaw the sweet corn kernels and green peas. Rinse tofu and dice it.
2. Heat stock in a pot. Add tofu, sweet corn kernels, green beans, and laver. Cook for 15 minutes. Season with sea salt and stir in caltrop starch slurry while cooking. Whisk the egg white and stir it in. Turn off the heat and cover the lid. Leave it briefly. Serve.

▌Indications and contraindications

This soup is velvety in texture. It is a great soup for those suffering from poor appetite and those who need pus draining after having surgery deep inside the neck or any operation in the abdomen or gastrointestinal system. Generally speaking, anyone may consume after any surgery. Only those with Asthenia or Coldness in the Spleen or Stomach meridians should consume in moderation.

Blended congee with dried scallops and dried tangerine peel

(makes 1 to 2 servings) Ref. p.124

▌Ingredients

3 dried scallops (steamed till soft)
1 small piece dried tangerine peel
60 g rice

▌Marinade for rice

oil
salt

▌Method

1. Soak dried tangerine peel in water till soft. Shred it. Break the dried scallops into fine shreds. Rinse the rice and drain. Add a dash of oil and a pinch of salt. Mix well.
2. Put all ingredients into a rice cooker. Add 6 bowls of water. Turn on rice cooker and let the congee programme completes. Use a blender or hand blender to puree the mixture. Serve.

▌Indications and contraindications

This congee is sweet in taste and is easy to digest. Those who suffer from compromised physical strength, poor appetite, restlessness, thirst, insomnia, or light sleep with many dreams after surgical operation would benefit from serving this congee. Generally speaking, everyone may consume it. Only gout patients should not eat the dried scallops.

▌Tips

After rinsing the rice, marinate it with some oil and salt before cooking. This makes the congee more velvety in texture. Before steaming the dried scallops, soak them in water until soft. Then steam them with the soaking water for 1 hour. You may then divide them into smaller portions and keep them in plastic bags. Keep in the freezer and thaw only what you need for that meal. The dried scallops would taste softer without any tough bit after steamed.

Perch and spinach thick soup

(makes 2 servings) Ref. p.126

▌ Ingredients

50 g spinach
100 g sea perch
100 ml evaporated milk
1 tbsp flour slurry

▌ Seasoning

oil
salt

▌ Method

1. Rinse the spinach and finely chop it. Steam the fish until done. Remove all bones.
2. In a pot, boil 3 bowls of water. Put in spinach and fish. Cook for 7 minutes. Add evaporated milk and seasoning. Stir in flour slurry. Cook until it thickens. Turn off the heat. Blend into a fine paste in a blender. Serve.

▌ Indications and contraindications

This soup is highly nutritious and is good for all ages. Those having difficulty or pain when passing stool, bloody stool, anaemia or dizziness after any bowel surgery would benefit from serving this soup. However, those taking blood thinner should avoid spinach as it may lead to haemorrhage.

Millet congee with lotus root starch

(makes 1 to 2 servings) Ref. p.129

▌ Ingredients

2 tbsp lotus root starch
10 dried Goji berries
60 g millet
10 g rice

▌ Seasoning

raw sugar

▌ Method

1. Rinse the millet and soak in water for 30 minutes. Rinse the rice and set aside. Add cold water to lotus root starch and mix into a smooth slurry. Rinse and soak Goji berries in water.
2. Boil 6 bowls of water. Put in the millet and rice together with the soaking water. Cook for 1 hour. Add Goji berries. Stir in lotus root starch slurry and sugar. Stir until sugar dissolves. Serve.

▌ Indications and contraindications

This congee is perfect for those who need semi-solid diet. It reduces the pain and chances of bleeding on surgical wounds. It also reduces the pain and bleeding when passing stool while promoting the recovery of the intestines. Those who had operations done to treat haemorrhoids, anal fistula or abscesses around the anus would especially benefit from serving this congee. Any patient may eat it after any surgery.

Rice noodles in fish broth with angled luffah and straw mushrooms

(makes 2 servings) Ref. p.132

▌Ingredients

1/2 angled luffah
4 straw mushrooms
4 g dried Goji berries
200 g fresh rice noodles
3 bowls fish bone broth

▌Seasoning

1/2 tsp sea salt

▌Method

1. Peel the angled luffah and cut into pieces. Rinse the straw mushrooms and cut each into halves along the length. Soak and rinse Goji berries in water.
2. Bring the fish broth to the boil. Put in angled luffah, straw mushrooms and Goji berries. Boil for 10 minutes. Put in the rice noodles and seasoning. Bring to the boil and serve.

▌Indications and contraindications

This noodle soup is tasty and flavourful. It is good for all ages. Those suffering from poor appetite, thirst, restlessness and feeling hot, thick phlegm in the throat, difficulty passing stool or urine after surgery would benefit from serving it. However, those with Coldness or Asthenia in the Spleen or Stomach should not consume.

Steamed rice with pumpkin and raisins

(makes 2 servings) Ref. p.134

▌Ingredients

150 g pumpkin
1 cup rice
2 tbsp raisins

▌Marinade for rice

oil
salt

▌Method

1. Peel and de-seed the pumpkin. Dice coarsely. Rinse the rice and drain. Add a pinch of salt and a dash of oil. Mix well.
2. Put rice and pumpkin into a rice cooker. Add 1 cup of water and turn on the cooker until the programme completes. Stir in raisins. Serve.

▌Indications and contraindications

This rice is sweet and tasty. It is good for all ages. Those suffering from Qi- and Blood-Asthenia, anaemia, pale complexion, dry throat, restlessness, thirst, difficulty passing stool or urine, or numbness due to Wind-Dampness after surgery would benefit from serving this rice. Generally speaking, all patients may consume after any surgical operation. Yet, those with Dampness accumulated in the body should eat in moderation.

Oatmeal with minced chicken and sweet corn kernels

(makes 1 to 2 servings) Ref. p.136

▌ Ingredients

50 g fresh chicken fillet
2 tbsp frozen sweet corn kernels
20 g rolled oats

▌ Marinade

1 egg white
sea salt
cornstarch

▌ Method

1. Finely chop the chicken. Add marinade and mix well. Leave it briefly. Thaw the sweet corn kernels.
2. Boil 2 bowls of water in a pot. Put in the sweet corn kernels and rolled oats. Cook for 5 minutes. Put in the chicken and stir to break it into bits. Cook till the chicken turns white. Turn off the heat and cover the lid. Leave it briefly. Serve.

▌ Indications and contraindications

This staple is nutritious and light. It is also easy to digest. It helps patients who can consume starchy food after surgical operations to recuperate. Generally speaking, any patient may consume after having surgery. Only those with kidney problems and those with high blood uric acid level should consume oatmeal in strict moderation.

Fried rice with dried scallops, egg white and grated ginger

(makes 2 servings) Ref. p.138

▌ Ingredients

3 to 4 dried scallops (steamed till soft)
2 egg whites
1 tbsp grated ginger
3 kale stems
2 bowls day-old rice
rice wine

▌ Seasoning

1 tsp sea salt

▌ Method

1. Break the dried scallops into fine shred after steamed. Whisk the egg whites. Rinse the kale stems and dice them.
2. Heat oil in a wok. Stir-fry ginger until fragrant. Add kale stems, dried scallops and rice. Toss till fragrant. Sizzle with wine and pour in egg whites. Toss until dry and each rice grain is heated through. Serve.

▌ Indications and contraindications

This rice is tasty and flavourful. It is good for all ages. Those suffering from general weakness, poor appetite, physical and mental exhaustion, restlessness and thirst would benefit from eating this rice. Any patient may consume after any surgery.

Fish ball ramen with lettuce

(makes 2 servings) Ref. p.140

Ingredients

50 g lettuce
150 g ramen
3 bowls fish bone broth
8 hand-beaten fish balls
1/2 tbsp shredded ginger

Seasoning

1/2 tsp sea salt

Method

1. Rinse the lettuce. Cook ramen in boiling water until just done. Drain and rinse in cold water. Drain again.
2. Boil the fish bone broth. Put in the fish balls, ginger and lettuce. Cook for 10 minutes. Season with sea salt and put in the ramen. Boil briefly. Serve.

Indications and contraindications

This noodle soup is tasty and flavourful. It is good for all ages. Those suffering from Dryness and Heat accumulated in the Stomach and intestines, restlessness, thirst, anaemia, headache, dry stool or constipation after surgical operation would benefit from serving this noodle soup. Generally speaking, everyone may consume.

Black glutinous rice congee with dried longans

(makes 2 servings) Ref. p.0142

Ingredients

10 dried longans
100 g black glutinous rice

Method

1. Rinse the dried longans. Rinse the black glutinous rice and soak in water.
2. Put dried longans and black glutinous rice into a pot. Add 6 bowls of water (including the soaking water for black glutinous rice). Boil for 1 hour. Serve.

Indications and contraindications

This congee is sticky and tasty. It helps alleviate low spirits, dry mouth and throat, physical and mental exhaustion, anaemia, sweating due to Asthenia, palpitations and insomnia among patients after surgery. Any patient may consume after any surgery.

Chicken congee with shiitake mushrooms

(makes 2 servings) Ref. p.144

▋ Ingredients

4 small dried shiitake mushrooms
1 organic chicken leg
60 g rice
1/2 tbsp shredded ginger
1/2 tbsp finely chopped spring onion

▋ Marinade

1/2 tsp sea salt
ground white pepper
1 tsp cornstarch

▋ Method

1. Soak mushrooms in water till soft. Cut off the stems and shred them. De-bone the chicken leg. Dice the chicken flesh. Add marinade to the chicken and mix well. Rinse the rice and add a dash of oil and a pinch of salt. Mix well.
2. In a pot, boil water with shiitake mushrooms, ginger, rice and chicken bones until the rice breaks down and the congee thickens. Remove the chicken bones. Put in the diced chicken. Cook for 5 to 6 more minutes. Sprinkle with spring onion. Serve.

▋ Indications and contraindications

This congee is velvety and tasty. Those who suffer from memory loss, insomnia, light sleep with many dreams, dizziness, physical and mental exhaustion after surgical operations would benefit from serving it. Generally speaking, all patients may consume after operations. Only gout patients should avoid eating mushrooms.

Stir-fried spaghetti with Iberian pork and beetroot

(makes 2 servings) Ref. p.146

▋ Ingredients

100 g beetroot
100 g Spanish Iberian pork
4 ears baby sweet corn
30 g white cabbage
150 g spaghetti

▋ Marinade for pork

1 tsp light soy sauce
1 tsp rice wine
cornstarch

▋ Seasoning

1/2 tsp salt
ground black pepper

▋ Method

1. Peel and rinse the beetroot. Finely shred it. Slice the pork thinly. Add marinade and mix well. Rinse the baby sweet corn. Cut each into halves along the length. Rinse the cabbage and shred it.
2. Boil a pot of water and add a pinch of salt. Put in the spaghetti and fan them out. Cook according to the instruction on the package. Drain and rinse in cold water. Drain again.
3. Heat some oil in a wok. Put in beetroot and spaghetti. Season with black pepper and salt. Toss until the spaghetti turns pink. Save on a serving plate.
4. Save some oil in the same wok. Heat it up and fry the pork until lightly browned. Add cabbage and baby sweet corn. Toss briefly. Transfer onto the serving plate around the spaghetti. Serve.

▋ Indications and contraindications

This noodle dish smells, looks and tastes divine. It is nutritious and delicious, good for all ages. It helps alleviate Qi- and Blood-Asthenia, pale complexion, nervous prostration, poor appetite and unstable blood pressure among cancer patients and those who had surgical operations. However, those taking blood thinning pills should consume beetroot with absolute care.

Lean pork soup with beetroot, tomatoes and potatoes

(makes 2 servings) Ref. p.152

▓ Ingredients

100 g beetroot
50 g tomatoes
50 g potatoes
150 g lean pork
2 slices ginger

▓ Seasoning

1/4 tsp sea salt

▓ Method

1. Peel beetroot, tomatoes and potatoes. Cut them into pieces. Slice pork and blanch in boiling water. Drain.
2. Put all ingredients into a pot. Add 6 bowls of water. Boil for 1 hour. Season with sea salt. Serve.

▓ Indications and contraindications

This soup is sweet and is good for all ages. Those suffering from Qi- and Blood-Asthenia, pale complexion, unstable blood pressure, poor appetite and slow peristalsis along the digestive tract after surgical operation would benefit from serving this soup. However, those with renal disease, diabetes and low blood pressure should use with care.

Double-steamed sea cucumber soup with walnuts and white fungus

(makes 2 servings) Ref. p.150

▓ Ingredients

20 g walnuts (shelled)
9 g white fungus
4 black dates
150 g pork shoulder blade bones
2 rehydrated sea cucumbers
3 slices ginger

▓ Seasoning

1/4 tsp sea salt

▓ Method

1. Rinse and soak walnuts and white fungus in water. Cut off the yellowish roots on the white fungus. Rinse the black dates. Blanch pork shoulder blade and sea cucumbers in boiling water. Drain.
2. Put all ingredients into a double-steaming pot. Pour in 3 bowls of boiling water. Double-steam for 3 hours. Season with sea salt. Serve.

▓ Indications and contraindications

This soup is nourishing and tasty. It is good for all ages. It helps alleviate symptoms among patients in the recuperation phase after surgery, such as Qi- and Blood-Asthenia, mental and physical exhaustion, insomnia, poor memory and frequent urination. Any patient may consume after surgery, but those suffering from gout should not eat the sea cucumbers.

Scorpionfish soup with chayote and tofu

(makes 2 servings) Ref. p.157

▌Ingredients

1 chayote
1 cube firm tofu
300 g weedy scorpionfish
3 g dried Goji berries
2 slices ginger

▌Seasoning

1/4 tsp sea salt

▌Method

1. Peel chayote and cut into pieces. Rinse and cut tofu into pieces. Dress the fish and rinse well. Wipe dry and fry in a little oil until lightly browned on all sides.
2. Boil water vigorously. Put in all ingredients and cook for 1 hour. Season with sea salt.

▌Indications and contraindications

This soup is flavourful and tasty. It is suitable for all ages. Patients suffering from poor Qi flow in the Liver meridian, anxiety, bloated abdomen and pain in the surgical wound when recovering from surgery would benefit from serving this soup. Women who gave birth by Caesarean section may feel pain in the wound and suffer from inflammation. After drinking this soup, they usually recover quickly. This soup is generally suitable for everyone.

Double-steamed silkie chicken soup with monkey-head mushrooms

(makes 2 servings) Ref. p.154

▌Ingredients

2 dried monkey-head mushrooms
6 red dates
2 dried scallops
1/2 organic silkie chicken
2 slices ginger

▌Seasoning

1/4 tsp sea salt

▌Method

1. Soak and rinse monkey-head mushrooms in water. De-seed the red dates. Soak dried scallops in water till soft. Dress the chicken and rinse well. Skin it and chop into pieces. Blanch in boiling water. Drain.
2. Put all ingredients into a double-steaming pot. Pour in 3 bowls of boiling water. Cover the lid and double-steam for 2 hours. Season with sea salt. Serve.

▌Indications and contraindications

This soup is fragrant and tasty. It is suitable for all ages. Those suffering from Blood-Asthenia, general weakness, physical and mental exhaustion after surgical operation on the digestive system would benefit from serving this soup. It is also recommend to those with poor appetite, mental exhaustion, thirst and dry mouth, "steaming bone" disorder and intermittent fevers. However, cancer patients with Dampness-Heat-Toxin accumulation should not consume.

Abalone soup with Goji berries and chrysanthemums

(makes 2 servings) Ref. p.160

▌ Ingredients

1 green-lip abalone
1 carrot
5 g dried Goji berries
6 g dried chrysanthemums
3 slices ginger

▌ Seasoning

1/4 tsp sea salt

▌ Method

1. Rinse Goji berries and chrysanthemums separately. Drain and set aside. Trim off the guts of the abalone. Scrub and rinse well. Blanch in boiling water briefly. Drain. Peel and cut carrot into chunks.
2. Boil abalone and carrot in 7 bowls of water for 2 hours. Add Goji berries and chrysanthemums. Boil for 5 minutes. Season with sea salt. Serve.

▌ Indications and contraindications

This soup is tasty and nourishing. It is good for all ages. Those suffering from red swollen eyes, poor eyesight, unstable blood pressure, frequent urinations at night, Qi-Asthenia, shortness of breath, high blood glucose level or cancer patients will benefit from serving this soup. However, those with high uric acid level in the blood and gout patients should not consume.

Double-steamed chicken soup with Huai Shan and fish maw

(makes 2 servings) Ref. p.164

▌ Ingredients

38 g Huai Shan (dried yam)
6 g dried Goji berries
100 g rehydrated fish maw
6 red dates
1 piece dried tangerine peel
1/2 free-range chicken

▌ Seasoning

1/4 tsp sea salt

▌ Method

1. Soak and rinse Huai Shan and Goji berries in water. Soak dried tangerine peel in water till soft. Drain. Blanch fish maw in boiling water briefly. Drain. De-seed the red dates. Chop chicken into big pieces. Blanch in boiling water. Drain.
2. Put all ingredients into a double-steaming pot. Pour in 3 bowls of boiling water. Cover the lid and double-steam for 3 hours. Season with sea salt. Serve.

▌ Indications and contraindications

This soup is flavourful and tasty. It is good for all ages. Those suffering from physical and nervous exhaustion, anaemia, pale complexion, soreness and weakness in the knees and lower back, or blood stasis when recovering from surgery would benefit from serving this soup. However, those with poor digestion should refrain from eating too much fish maw, as it is hard to digest.

Lean pork soup with tea tree mushrooms, water chestnuts and carrot

(makes 2 servings) Ref. p.166

▌Ingredients

20 g tea tree mushrooms
6 water chestnuts
10 g almonds
1 carrot
150 g lean pork

▌Seasoning

1/4 tsp sea salt

▌Method

1. Rinse and soak the tea tree mushrooms in water. Cut off the base of the stems. Peel water chestnuts and carrot. Then cut into pieces. Rinse the almonds. Slice the pork and blanch in boiling water. Drain.
2. Put all ingredients into a pot. Add 1 litre of water. Boil for 1 1/2 hours. Season with sea salt. Serve.

▌Indications and contraindications

This soup is tasty and sweet. It is good for all ages. Patients suffering from Kidney-Asthenia, difficulty in urinating, frequent urination, oedema, insufficient Qi in Lung meridians, shortness of breath or much phlegm while recovering from surgery would benefit from serving this soup. It also helps cancer patients recover. However, those allergic to mushrooms should avoid.

Partridge soup with gingkoes and lily bulbs

(makes 2 servings) Ref. p.168

▌Ingredients

15 gingkoes
30 g dried lily bulbs
30 g soybeans
3 slices ginger
4 red dates
1 partridge

▌Seasoning

1/4 tsp sea salt

▌Method

1. Shell and core the gingkoes. Rinse well. Soak and rinse lily bulbs and soybeans in water separately. De-seed the red dates. Dress the partridge and rinse well. Blanch partridge in boiling water briefly. Drain.
2. Put all ingredients into a pot. Add 7 bowls of water. Boil for 2 hours. Season with sea salt. Serve.

▌Indications and contraindications

This soup is nourishing and delicious. It is good for all ages. Those suffering from insufficient Qi (vital energy) in the Lung meridian, cough, shortness of breath, frequent urinations at night, nervous prostration, insomnia, or light sleep with many dreams while recovering from surgery would benefit from serving this soup. Patients may feel free to consume after any surgical operation. Yet, gout patients should not eat the soybeans.

Luo Han Guo tea with green olives

(makes 2 servings) Ref. p.171

▌Ingredients

8 green olives
1/2 golden Luo Han Guo (freeze-dried)

▌Method

1. Rinse the olives. Crush gently with the back of a knife. Finely chop the Luo Han Guo.
2. Put all ingredients into a pot. Add 5 bowls of water. Bring to the boil and keep cooking over high heat for 30 minutes. Serve.

▌Indications and contraindications

This tea is sweet in taste and is suitable for all ages. It helps alleviate dry mouth and throat, bitterness in the mouth, difficulty in swallowing and constipation among cancer patients after radiation or chemotherapy. However, those with Coldness in the Spleen and Stomach should add dried tangerine peel and fresh ginger.

Job's tears tea with sugarcane and lalang grass rhizome

(makes 2 servings) Ref. p.174

▌Ingredients

120 g sugarcane
120 g fresh lalang grass rhizome
30 g raw Job's tears

▌Method

1. Rinse the sugarcane. Cut each into half along the length. Rinse lalang grass rhizome. Cut into short lengths. Soak and rinse the Job's tears in water.
2. Boil 6 bowls of water. Put all ingredients in. Bring to the boil again over high heat. Turn to medium heat and cook for 40 minutes until the liquid reduces to 2 to 3 bowls. Serve.

▌Indications and contraindications

This tea is sweet in taste. Those cancer patients with dry throat and mouth, restlessness and bitterness in the mouth would benefit from it. It also helps alleviate side effects of radiotherapy such as sore throat ulcers and difficulty in swallowing. Leukaemia patients with Heat-related symptoms suffering from haemorrhage can also benefit from this tea. However, those with Asthenic body type and frequent urination at night should not consume. Pregnant women should skip the Job's tears.

Grape and lotus root juice

(makes 2 servings) Ref. p.176

Ingredients

150 g freshly squeezed grape juice
200 ml freshly squeezed lotus root juice

Method

1. Rinse fresh grapes in water. Squeeze into juice in a juicer with the skin and seeds. Peel the lotus root and rinse well. Squeeze into juice.
2. Put grape juice and lotus root juice into a pot. Add 100 ml of water. Bring to the boil over low heat. Let cool before serving.

Indications and contraindications

This drink is light and nourishing. It is suitable for all ages and generally good for everyone. It helps alleviate haemorrhage, stomach ulcer, poor appetite, anaemia and general weakness among cancer patients after radiation or chemotherapy. Women with ovarian cancer may feel free to consume frequently.

Lean pork soup with dried Bok Choy, carrot and dried duck gizzards

(Makes 2 to 3 servings) Ref. p.178

Ingredients

60 g dried Bok Choy
1 carrot
100 g lean pork
2 dried duck gizzards
2 candied dates

Seasoning

1/4 tsp sea salt

Method

1. Rinse and soak the dried Bok Choy in water. Cut into short lengths. Peel and cut carrot into pieces. Rinse the pork and slice it. Blanch pork and dried duck gizzards in boiling water briefly. Drain.
2. Put all ingredients into a pot. Add 1 litre of water. Boil for 1 1/2 hours. Season with sea salt. Serve.

Indications and contraindications

This soup is tasty. It is suitable for all ages. Stomach cancer patients with insufficient Yin in the Stomach would benefit from it. It also helps alleviate certain side effects of radiation or chemotherapy, such as dry mouth, thirst, weight loss, poor appetite and indigestion. Any cancer patient may consume.

Lean pork soup with mushrooms and Goji berries

(Makes 2 to 3 servings) Ref. p.180

▓ Ingredients

1 head dried maitake mushrooms
4 dried shiitake mushrooms
3 g dried Goji berries
6 red dates
1 piece dried tangerine peel
200 g lean pork

▓ Seasoning

1/4 tsp sea salt

▓ Method

1. Rinse maitake and shiitake mushrooms. Cut off the stems of the shiitake. Soak and rinse dried tangerine peel and Goji berries in water. De-seed the red dates. Slice the pork and blanch in boiling water. Drain.
2. Put all ingredients into a pot and add 1 litre of water. Bring to the boil and cook for 1 hour. Season with sea salt. Serve.

▓ Indications and contraindications

This soup is fragrant and delicious. It is good for all ages. It helps suppress the growth of cancerous cells, alleviates side effects of radiotherapy and chemotherapy, and boosts immune system. However, those allergic to mushrooms should not consume.

Congee with shredded pork and fish maw

(makes 2 to 3 servings) Ref. p.182

▓ Ingredients

50 g pork shoulder butt
30 g rehydrated fish maw
60 g rice
2 olices ginger
1 small bunch chopped spring onion

▓ Marinade

salt
ground white pepper
caltrop starch

▓ Seasoning

1/4 tsp sea salt

▓ Method

1. Shred the pork. Add marinade and mix well. Leave it for 30 minutes. Shred the fish maw. Rinse the rice. Add a dash of oil and a pinch of salt. Mix well.
2. Boil 6 bowls of water in a pot. Add rice and ginger. Cook until the rice breaks down and to your desired consistency. Add shredded pork and fish maw. Bring to the boil. Season with sea salt. Sprinkle with spring onion on top. Serve.

▓ Indications and contraindications

This congee is velvety smooth and very nourishing. It is good for all ages. Women suffering from gynaecological cancers (such as cervical or ovarian cancer) would benefit from it. It also helps alleviate weight loss, general weakness and poor appetite after cancer removal surgery, during and after radiation or chemotherapy. However, those with Dampness in the Spleen and excessive phlegm should not consume.

Whitebait congee with dried scallops

(makes 2 servings) Ref. p.184

Ingredients

200 g fresh whitebaits
2 to 3 dried scallops
60 g rice
1 tbsp finely shredded ginger
1 tbsp finely chopped spring onion

Seasoning

1/4 tsp sea salt

Method

1. Rinse the whitebaits. Drain. Soak dried scallops in water till soft. Break into fine shreds. Rinse rice. Add a dash of oil and a pinch of salt. Mix well.
2. Pour 6 bowls of water in a pot. Add rice, dried scallops and shredded ginger. Cook until the rice breaks down and to your desired consistency. Add whitebaits and sea salt. Bring to the boil. Sprinkle with spring onion. Serve.

Indications and contraindications

This congee is velvety smooth and delicious. It is good for all ages. Stomach and lung cancer patients with Yin-Asthenia and insufficient body fluid secretion would benefit most. It also helps alleviates the side effects during or after radiotherapy, such as accumulated Heat, Yin-Asthenia, thirst, restlessness, weight loss, poor appetite, "steaming bone" disorder with intermittent fever, and dry cough without phlegm. However, those suffering from gout should not consume.

Dace ball soup with tomato and tofu

(makes 2 to 3 servings) Ref. p.186

Ingredients

120 g tomatoes
1 cube tofu
100 g minced dace
1 sprig coriander

Seasoning

1/4 tsp sea salt

Method

1. Peel the tomatoes and cut into pieces. Rinse tofu with water and cut into pieces. Shape the minced dace into balls. Cut off the roots of coriander. Rinse and cut into short lengths.
2. Boil 4 bowls of water. Put in tomatoes and tofu. Bring to the boil. Add dace balls. Boil for 10 minutes. Season with sea salt. Sprinkle with coriander. Serve.

Indications and contraindications

This soup is fragrant and appetizing. It is suitable for all ages. It helps alleviates side effects associated with radiation and chemotherapy for cancers, such as insufficient body fluid, poor appetite and thirst. Add a dash of fresh ginger juice or ground white pepper for patients who also suffer from vomiting. It is generally good for everyone. Only those suffering from gout should consume tofu in strict moderation.

手術前後飲食自療
養護好體質

Pre- / Post-operative healthful recipes
All-round wellness

作者	Author
芳姐	Cheung Pui Fong

策劃/編輯 　　Project Editor
譚麗琴 　　Catherine Tam

攝影 　　Photographer
細權 　　Leung Sai Kuen

美術設計 　　Art Design
陳玉菁 　　Ceci Chen

出版者 　　Publisher

Forms Kitchen

香港鰂魚涌英皇道1065號
東達中心1305室

Room 1305, Eastern Centre, 1065 King's Road,
Quarry Bay, Hong Kong.

電話 　　Tel: 2564 7511
傳真 　　Fax: 2565 5539
電郵 　　Email: info@wanlibk.com
網址 　　Web Site: http://www.wanlibk.com
　　　　　　http://www.facebook.com/wanlibk

發行者 　　Distributor

香港聯合書刊物流有限公司

香港新界大埔汀麗路36號
中華商務印刷大廈3字樓

SUP Publishing Logistics (HK) Ltd
3/F., C&C Building, 36 Ting Lai Road,
Tai Po, N.T., Hong Kong

電話 　　Tel: 2150 2100
傳真 　　Fax: 2407 3062
電郵 　　Email: info@suplogistics.com.hk

承印者 　　Printer

Best motion

出版日期 　　Publishing Date
二零一七年十二月第一次印刷 　　First print in December 2017